······The Ultimate Guide to······

Smoking Meat, Fish, and Game

The Ultimate Guide to

Smoking Meat, Fish, and Game

HOW TO MAKE EVERYTHING FROM DELICIOUS MEALS TO TASTY TREATS

Monte Burch

Skyhorse Publishing

Skyhorse Publishing books may be purchased in bulk at special discounts for sales promotion, corporate gifts, fund-raising, or educational purposes. Special editions can also be created to specifications. For details, contact the Special Sales Department, Skyhorse Publishing, 307 West 36th Street, 11th Floor, New York, NY 10018 or info@skyhorsepublishing.com.

Skyhorse® and Skyhorse Publishing® are registered trademarks of Skyhorse Publishing, Inc.®, a Delaware corporation.

Visit our website at www.skyhorsepublishing.com.

10 9 8 7 6 5 4 3 2 1

Library of Congress Cataloging-in-Publication Data is available on file.

Cover design by Brian Peterson
Cover photo credit Thinkstock

Print ISBN: 978-1-63220-471-4

Ebook ISBN: 978-1-63220-792-0

Printed in China

Contents

PART 1:

Smoking

①

All About Smoking and Salt Curing

SMOKING, COMBINED WITH using salt to preserve meats, is one of mankind's oldest and most important survival skills. The use of salt has a long history. Salt is one of the most important elements and was in use long before recorded history. Since the dawn of time, animals have instinctively forged trails to natural sources to satisfy their need for salt. In turn, ancient man obtained his salt from eating animal meat. As he turned to agriculture and his diet changed, he found that salt (maybe seawater) gave other foods the same salty flavor he was accustomed to in meat. Over many millennia, man learned how salt helped with preserving food. Salt was also an important trade item. Many nomadic bands carried salt with them and traded it for other goods. Wars were even fought over salt, according to the Salt Manufacturers Association (www.saltinfo.com).

About 4,700 years ago, the Chinese Peng-tzao-kan-mu, one of the earliest known writings, recorded more than 40 types of salt. It described two methods of extracting and processing salt, similar to methods still in use. Writings on salt no doubt also existed on ancient clay tablets and on Egyptian papyri. Even without written evidence, we can be fairly certain that salt-making and use was a feature of life in all ancient communities.

For centuries, salt has been used to preserve foods such as meat, fish and dairy products. Even with the development of refrigeration, salt preserving remains an important aid to food hygiene. Salt acts as a binder as it helps extract the myofibrillar proteins in processed and formed meats, binding the meat together and reducing cooking losses. Used with sugar and nitrate, salt gives processed meats, such as ham, bacon and hot dogs, a more attractive color.

In addition, the use of smoke, both as a preservative and during cooking, has also been—and still is—very important. The first smoking, or "cooking," was probably done over a smudge, a smoky fire in a cave. Today smoking and smoke-cooking meats can be just about as primitive, or extremely sophisticated.

Several types of smoking can be done depending mostly on the temperature of the smoke heat applied. This includes dry or cold smoking, hot smoking, and barbecuing. Dry or cold smoking is used to impart the flavor of the smoke as well as help to dry out the meat. This is used for drying jerky and sausages, as well as for drying and adding flavor to hams, bacon, and other meats. Temperatures normally are not allowed to get above 100°F. Cold smoking is also quite often used in conjunction with the brining or salt curing of many different types of meats and meat products. All cold-smoked meats must be further cooked before consumption.

Hot smoking utilizes smoke and heat to add flavor at the same time that the meats are cooked. This type of smoking is done indirectly over hot coals, or with gas or electric heat, and temperatures rarely exceed 250°F. Cooking time is several hours. This is a common method for cooking briskets, ribs, and other popular smoked foods. Hot smoking may or may not involve water, and may be dry or moist cooking.

One of the original forms of smoke cooking is the old-time pit barbecue, and some believe the word *barbecue* comes from the Caribbean word *barbacoa*, translated as "sacred fire pit." Outdoor cooking on the barbecue grill is a very popular pastime and a favored method of cooking. In the South, *barbecue* usually refers to roast pork; in the Southwest, it usually refers to beef. Barbecuing is cooking with direct heat and may or may not include smoking. Smoking does, however, impart more flavor to barbecued foods.

In addition to salt curing, I have included both cold-smoking and hot-smoking methods in this book. I have also included some barbecuing methods utilizing smoke, although there are literally thousands of recipes and entire books written on barbecue cooking.

I'm not sure that my grandparents would have called salting and smoking hams and bacon a joy, as it was a necessity of survival for them and their parents before the advent of refrigeration. I can, however, remember the fun they all had during the family and community hog butchering days. Their skills were passed down from generations, but the skills are not hard to learn. Using smoke and salt to cure meats for preservation is a fun, easily learned process as well as a great way of serving your family great-tasting foods.

Other than cold-smoking hams and bacon, there was no other type of smoking in our family when I was growing up. Later, once I discovered that our home-cured hams were great slow-cooked over a rotisserie, I discovered the joy of smoking different foods, and I've now been smoking meats for forty years or more. I have smoked everything from salmon to whole hogs, with wonderful results. Which is why it's no wonder the process of smoking in all forms has become increasingly popular.

I hope this book brings you the joy of smoking.

THE MOST IMPORTANT tools for smoking, of course, are smokers. In days past, the smokehouse was a traditional building found on most farmsteads. These buildings often served two purposes—as a place to cure, age, and then store hams, shoulders, and other meats, as well as a place to cold-smoke them. In cold climates, the meats were usually left hanging in the smokehouse until consumed. These smokehouses varied greatly in design and size. Some were simply slats fastened to a frame; others were made of chinked

② Smokers and Other Tools

logs. Some were more elaborate and constructed to match other farm outbuildings. All, however, had a system of venting out the smoke, and many had screens over the openings to keep out pests. Sizes ranged from just a few square feet, with smoke generated in a fire outside and piped in, to larger structures capable of smoking larger numbers of hams, shoulders, and bacon. The smoke fire was built inside, either on the ground or in a stone fire pit. My granddad had such a smokehouse—one large enough to smoke the meat from a

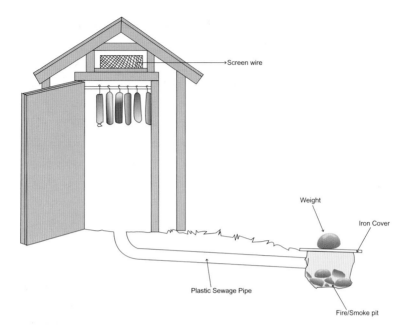

Screen wire

Weight

Iron Cover

Plastic Sewage Pipe

Fire/Smoke pit

ABOVE: Smoking requires smokers, and if you intend to smoke in volume, you may wish to make up your own traditional wooden smokehouse.

dozen or so hogs, as were often butchered during family and community hog butchering days. Even after he hadn't smoked in the building for years, it still held the smell of hickory smoke.

Homemade Smokers

You can build your own wooden-frame traditional smokehouse, although these days, most of us don't have the space or the need for a traditional smoke building. A better choice today is a smaller smoker. You can also make your own cold smoker quite easily. Many years ago, when I first became interested in cold smoking, I constructed my first smoker from plywood—a simple box with a door on one side, a hole on the top, and an electric hot plate holding a pan of soaked wood chips. I used the box for cold-smoking salmon and other fish, as well as some bacon.

My next smoker consisted of a fifty-gallon metal drum. I cleaned the drum thoroughly and cut out both ends. I dug a three-foot-deep by three-foot-wide hole for a fire pit, then dug a six-inch trench twelve feet long from the pit to the barrel.

It's important that the trench run slightly uphill to allow the heat-driven smoke to rise up and into the smoker barrel. I used rocks to line the fire pit, installed some six-inch stovepipe in the trench, and covered the pipe with soil. I then set the barrel in place over the end of the stovepipe. I simply placed a couple of broom-handle sticks across the barrel top, hung hams and bacon from the broomsticks, and placed the cut-off barrel top over the broom handles. I built fire in the pit, placed a metal sheet over it, and regulated the airflow by propping the metal sheet cover up or down with a rock. I placed a tarp over the barrel to create a primitive damper and to regulate smoke. This simple smoker could cold-smoke a couple of hams quite easily.

A few years later, I improved my smoker using a discarded refrigerator. As I experimented, I discovered that a refrigerator smoker can be "fired" in two ways. The first and original method was to cut a hole in the bottom or lower side of the refrigerator and install a metal stovepipe. I set this over my original stovepipe and fire pit arrangement. I

BELOW: A cold smoker can easily be made using a barrel or an old recycled refrigerator.

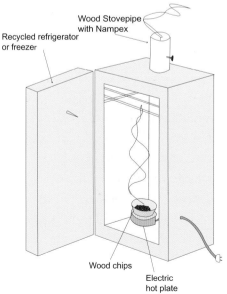

cut a smoke opening in the top of the refrigerator and added a damper, a simple metal plate held in place with one screw. A wooden stovepipe with damper could also be used. Lastly, I drilled a hole near the top of the refrigerator and installed a dial meat thermometer with silicone caulking applied around it.

The only problem with this style of smoker, as well as the smokehouses of old, is it requires a lot of attention and effort to maintain the fire for proper cold smoking. A number of years later, I added an electric hot plate to the bottom of the refrigerator. Placing a pan of water-soaked wood chips on the hot plate made smoking easier, but I still had to monitor the thermometer and refill the chip pan.

You can also make a propane-fueled smoker using a large barrel. This smoker can be used for both cold and hot smoking. Again the bottom is cut out and the top lid removed. Fire is provided by a cast-iron fish or turkey fryer with burner regulator.

BELOW: You can also make up your own propane-powered hot smoker using a barrel and replacement fish-fry burner and regulator.

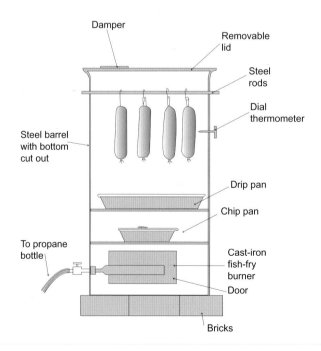

You will, of course, need racks to hold a wood chip pan, a water/ grease pan, and for the meat. A dial thermometer, or a remote probe thermometer, and a damper should be placed in the top lid.

Purchased Smokers

These days smoking is extremely popular, and a number of manufactured smokers are available. They come in a wide range of sizes, styles,

BELOW: A number of manufactured smokers are available in a wide range of types and sizes, fueled by electricity, charcoal, or wood.

prices, and types of fuel used. Fuels include charcoal, wood, wood pellets, gas, and electric. The simplest to use are the electric smokers; simply plug them in and add wood chips and meat. Gas is also easy; just add wood chips and meat. The newer pellet smokers are extremely easy. Set to the desired temperature and fill the hopper with the flavor or variety of wood pellets desired.

BELOW: One of the simplest and longtime traditional smokers is the domed-lid water smoker.

Charcoal-fueled smokers require a bit more effort as well as more attention, but many prefer the taste of charcoal-smoked meats. Purist smoker chefs prefer dedicated wood smokers, but these require the most work and monitoring.

Styles include simple, water-pan, barbecue-style units; the larger dedicated indirect-heat smokers; electric-fired and gas-fired smoker/cookers; and a number of barbecue-grill smokers. Prices range from around $50 for simple domed units to several thousand dollars for smokers big enough to do a whole hog. Your first step is to determine what you want to smoke and what type of smoking you intend to do. Do you want a dedicated smoker? Do you want the convenience and ease of gas or electric, or do you prefer to smoke with charcoal or wood? How much do you want to spend? How much will you smoke at a time? Do you want to both smoke and grill? Where will you store the smoker?

The simplest and most economical are the dome-lid, water-pan smokers. Over the years, I've tested a number of these, including the Bass Pro Shops/Brinkmann Smoke 'n Grill, which you can find along with other brands and types of smokers at Bass Pro Shops and other retailers that sell smoking and barbecuing supplies. The Smoke 'n Grill is a charcoal smoker, and the style has been around for a long time.

They're economical to buy and run, and they can be used to either hot-smoke or grill. The model I've tested will grill or smoke up to fifty pounds of meat. Features include easy-to-clean porcelain-coated charcoal and water pans; a large front door for adding charcoal, wood chips, and water; two cooking racks; and a temperature gauge. Smoke cooking with water is extremely easy to do and offers a pleasant surprise if you haven't tried it. This unique cooking process combines aromatic smoke and steam to continually baste the food while the indirect heat slowly cooks it, creating delicious, succulent smoked meals. Although the unit I tested is charcoal fired, gas-fueled water cookers are also available. Most of these units can also double as barbecue grills.

These charcoal-fueled smokers utilize a pan holding water or marinades over the coals, and between the meat, providing a semi-indirect heat for long smoking times.

ABOVE: An electric smoker, such as the Bradley smoker, provides almost hassle-free smoking. Electric smokers are easy to operate and provide steady, even heat at the temperature desired, from cold to hot smoking.

Another choice is an electric smoker. These units can be used for both cold and hot smoking. Most units will run from a little over $200 to around $600. I've tested two, the Bradley and the Masterbuilt Digital models. They're very easy to operate, and they maintain a constant regulated heat. Turn them on, set to the heat desired, add the smoke materials, fill with meat, and then leave them alone. Most also have provisions for a water pan to add moistness to the cooking.

One such unit I've used for years is the Bradley smoker. Although mine is an older model, their new digital models offer a number of advantages. All models will hot-smoke at up to 320°F and cold-smoke, slow-roast, or dehydrate at 140°F.

BELOW: Bradley smokers utilize Bradley pressed-wood bisquettes to provide the smoke.

ABOVE: The smoke generator feeds bisquettes at a precise burn time of 20 minutes per bisquette.

Merely turn the thermostat to the desired temperature. One of the unique features of the units is they burn Bradley flavored bisquettes, preformed wood chip discs that self-load into the smoker. Precise burn time is twenty minutes per bisquette. The insulated smoking cabinet is easy to use any time of the year, as I've discovered many a winter month. Six removable racks accommodate large loads and allow heat and smoke to circulate evenly. The Bradley has an easy, front-loading design. The Bradley unit also has a separate smoke generator accessory unit that fastens to the cabinet. The smoke generator can be removed and the accessory pipe used to set the smoke generator a distance from the smoker unit. Using only the smoke generator, you can create a true cold smoke at 80°F to 100°F for drying.

Probably the most versatile electric smoker I've tested is the Masterbuilt Electric Digital Smoker from Bass Pro and other retailers. You can do anything, from simple barbecuing to cooking sausage and smoking all types of meat. The digitally controlled electric smoker comes with a remote meat probe, inside light, and a glass door.

BELOW: An accessory for the Bradley smoker sets the smoke generator a distance from the smoker, allowing for true cold smoking at 100°F or less.

ABOVE: The Masterbuilt Electric Digital Smoker is an extremely versatile electric smoker for everything from cold to hot smoking and barbecue cooking.

You can actually watch your meat cook and smoke without opening the door. The push-button digital temperature and time control makes smokehouse cooking as easy as grilling. The thermostat-controlled temperature creates even, consistent smoking from 100°F to 275°F. You can easily cold-smoke or hot-smoke. The built-in meat probe helps ensure perfectly cooked food every time and instantly reads the internal temperature of the meat. Wood chips are loaded through a convenient door in the side. The Masterbuilt smoker also features a drip pan and a rear-mounted grease pan for easy cleanup. The removable pan keeps food moist. The four removable, chrome-plated cooking racks have 720.5 square inches of cooking space for large quantities of food. An air damper controls the smoke.

For true hot-smoked, barbecue-style smoking, nothing beats a dedicated indirect smoker. The better units can also be used as cold

BELOW: The digital controls make setting and regulating heat and time quick, easy, and precise.

ABOVE: The Masterbuilt features a removable pan for easy loading of wood chips for smoking.

BELOW: The Masterbuilt water pan provides for water smoking for hot-smoking moistness.

ABOVE: A built-in remote meat probe in the Masterbuilt provides an instant digital readout of internal meat temperatures.

BELOW: An external drip tray makes cleanup easy.

ABOVE: The glass door on the Masterbuilt allows you to see the meat as it smokes.

ABOVE: A built-in light on the Masterbuilt makes nighttime smoking easy.

smokers. These are fueled by charcoal or wood, some newer ones by pellets. They also range in size from tailgater models to huge smokers capable of holding a whole hog. A catering friend of mine has a custom-made unit, on a trailer, that can smoke-cook three hogs, or a whole steer. Prices for these types of smokers begin at $600 to $700. I've also tested a number of these, wearing out a couple over the years. This is one type of smoker you do not want to scrimp on. Cheaper units are made of thinner metal, are less rigid, and simply won't last as long, as I've found out the hard way. These units, however, even the smaller ones, do require some space and are fairly heavy. They should be on a noncombustible surface such as concrete or gravel.

The 16" Horizon Classic Smoker from Bass Pro and other retailers is a fairly common size for backyard smokers, although larger units are available. The unit has the offset firebox for indirect cooking and a large cooking area that can easily smoke enough food for sixty people. Most importantly, the unit is welded from new structural steel with quarter-inch pipe. The company offers a lifetime warranty against burnout. The hinged lids are also heavy-duty quarter-inch piping. A 16" x 16" firebox provides long cooking times and fuel efficiency, and the unit will burn wood, charcoal, or a combination of both. The

BELOW: An indirect, dedicated smoker, such as the Horizon, is the ultimate for serious smokers. Dedicated smokers can cold-smoke, hot-smoke, even barbecue, and they have the choice for long, low temperature or hot smoking.

ABOVE: The big firebox on the Horizon, separate from the meat-cooking area, provides plenty of space for loading lots of fuel for long cooking times.

BELOW: The Horizon smokestack with damper provides for easy regulation of heat and smoke.

ABOVE: A dial thermometer provides temperature information for easy control.

BELOW: The big cooking area on the Horizon provides lots of space for smoking for a crowd and smoking larger cuts of meat.

16" x 32" horizontal cooking chamber has 680 square inches of available cooking surface. The side firebox door provides easy access for ash removal and damper control. The smokestack also has a damper control. A wood storage shelf is under the cooking chamber. A grease drain and a cleanout hook make cleanup hassle-free and prevent flare-ups. Twelve-inch wagon-style wheels make the 300-pound unit easy to move around. Chrome spring handles stay cool during smoker use and last for years. A thermometer in a threaded port indicates temperature and cold- and hot-smoking ranges.

The standard and very popular barbecue grills can also be used for hot smoking. These are available in a wide range of sizes, styles,

BELOW: A standard barbecue grill, if it is large and covered, can be used for smoking. The grill must also have a temperature gauge rather than a heat range gauge.

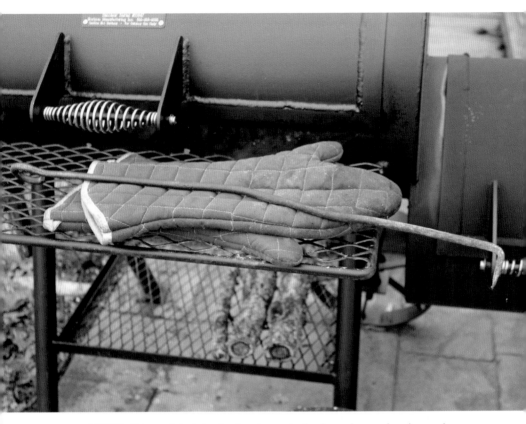

ABOVE: Other tools include a fireplace-type poker for charcoal or wood smokers, and heavy-duty insulated mitts.

and prices from tabletop to huge built-in models. Some models can also be used for cold smoking. Models include both charcoal- and gas-fired. The larger units, or those with larger cooking surfaces, are best for smoke cooking because they allow for more temperature control and indirect cooking methods.

Salt-Curing Tools

Salt-curing meats also requires some tools; the amount, number, and types depending on what kind of salt curing and smoking you do and how involved you become. If you do your own butchering, you'll need tools for those chores. Regardless, you'll need sharp knives for cutting up meat and sharpeners for keeping the knives sharp. Meat-processing or butcher knives come in a wide variety of sizes and shapes, with different types used for different chores. Wide-bladed knives are best for slicing meat into small chunks or slices. The more flexible thin-bladed knives are best for deboning meat.

I prefer the rounded-tip butcher knives for cutting meat into smaller pieces. The upswept tip doesn't "catch" on meat as you slice it. Butcher knives can be purchased separately or in sets. Some sets even include a sharpening steel as well. For instance, the RedHead Deluxe Butcher Knives Kit includes paring, boning, and butcher knives; a meat cleaver; spring shears; a square-tube saw; a honing steel; a cutting board; butcher's apron; and six pairs of gloves—all in a case.

Keeping knife blades sharp is an extremely important facet of any type of meat preparation. This is a continuous operation, and using the proper tools can make it easy to have sharp knives as needed. A wide variety of knife-sharpening devices is available, ranging from the simple but extremely effective butcher's steel to powered sharpeners. The powered hones, such as those from Chef'sChoice, can make the chore of sharpening a dull knife quick and easy.

BELOW: You'll also need knives for cutting and processing meats. A variety of boning and butchering knives is a great help.

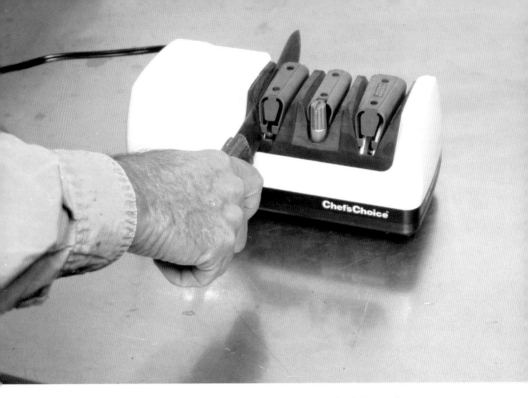

ABOVE: Keeping knives sharp is extremely important. The Chef'sChoice electric sharpeners are great for putting on a good edge and keeping knives sharp.

BELOW: A good sharpening steel, kept by the worktable, can be used for quick touch-ups.

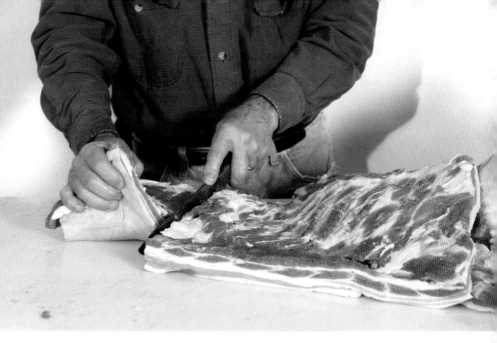

ABOVE: You'll need a good, solid, easily cleaned work surface. A large nonporous cutting board is necessary.

You'll also need a work surface. It must be a good, solid surface that is easily cleaned. Salt-curing large meat cuts, such as hams and shoulders, takes quite a bit of space. Your kitchen table or kitchen countertop will work, but it should not be wood, or even the traditional butcher block. The surface should be nonporous and easy to clean and disinfect. I remember my mom using a well-kept and well-cleaned piece of "oilcloth" spread over an outside table for the chore. A very large synthetic cutting board can also be used. You'll need smaller cutting boards for cutting up the meat as well.

If you're making sausage, you'll need a meat grinder. This can be a hand-cranked or powered meat grinder. If you're doing only small batches, a hand grinder will suffice. This is economical and easy to use. Of course, hand grinders grind only as fast as your muscle power can hand-crank. With larger amounts of meat, you'll really appreciate a powered meat grinder. These are available in a number of sizes, depending on horsepower, which determines the amount of meat that can be ground in an hour.

Smaller models will do the job, but with less power, they are slower. Top models can grind up to 720 pounds per hour. The Bass Pro

ABOVE: If you intend to make sausage, you'll need a meat grinder. A top-quality, heavy-duty grinder such as the LEM electric grinder is the ultimate for the home meat processor.

Shops Electric Meat Grinders by LEM Products are some of the best on the market. They all feature a full two-year warranty and have heavy-duty construction featuring stainless steel, easily cleaned housing, grinder head and auger, a permanently lubricated motor, a built-in circuit breaker, all-metal gears with roller bearings, a heavy-duty handle, and 110 V power. Standard accessories for all models include a large-capacity meat pan, a meat stomper, a stainless steel grinding knife, stainless steel stuffing plates, one stainless steel 3/16" Fine plate and one stainless steel 3/8" Coarse plate, and three stuffing tubes. Four models are available from Bass Pro. The model I tested is the no. 12, featuring a .75-horsepower motor and capable of grinding 360 pounds per hour. Sausage stuffing is what makes sausage, and a grinder with a stuffing tube makes the chore easier. For larger volumes of meat, and when using smaller-diameter stuffing tubes, you might want to go for a vertical stuffer.

You'll need a kitchen scale to weigh exact amounts of meat: 22- and 44-pound models are available at Bass Pro. The 22-pound model comes with a stainless steel tray. Other miscellaneous items include latex gloves, measuring cups and spoons, glass bowls, and a meat

ABOVE: A sausage stuffer, such as the LEM vertical model, makes sausage stuffing quick and easy.

thermometer. A digital model with a separate temperature probe is ideal. The temperature probe can stay in the meat while the meat is in the smoker. You'll also need tongs, barbecue mitts, and lots of food-grade plastic containers and bags.

BELOW: An electric meat slicer, like one of the many models available from Chef'sChoice can make quick work of slicing bacon and corned beef.

AS WITH ALL types of food preparation, safety is extremely important in smoking and salt-curing meats, which requires precautions in handling meats, in preparation areas, and in storing both fresh and processed meats. The use of safe and high-quality meats is also important. Don't, however, let this take the joy out of smoking and salt curing. These processes have been in use for ages. Just follow common sense safety rules.

3

Selecting Meats and Meat Safety

Safe Meat

The first prerequisite is safe meat. Use only meat from disease-free animals, poultry, and fish. It's important to be aware of diseases, such as CWD (chronic wasting disease) found in wild deer, elk, and moose, although there is no current evidence that CWD is transmittable to humans. Health officials, however, recommend that human exposure to the CWD agent be avoided. Hunters in CWD areas are advised to completely bone out harvested cervids in the field and not consume possibly infected parts, such as the brain, eyes, spinal cord, lymph nodes, tonsils, and spleen. Do not shoot, handle, or consume any animal that is acting abnormally or appears sick. They should also take simple precautions when field-dressing, including wearing field-dressing gloves.

Trichinosis, or trichinellosis, is a disease caused by a parasite called *Trichinella spiralis*. It's important to avoid eating the under-cooked meat of pork, bear, cougar, wild boar, and walrus. Make sure the meat of these animals is cooked to an internal temperature of 160°F before consumption. Poultry should be cooked to an internal temperature of 165°F and fish to 160°F.

Meat that is tainted or unsafely butchered or cut up also poses serious health problems, especially from *Escherichia coli* bacteria. Food

poisoning by *E. coli* is a serious and deadly problem that can make you extremely ill, even kill you. Many strains of *E. coli* exist; most are normal inhabitants of the small intestines and the colon, and are non-pathogenic. *E. coli* 0157:H7, however, is a dangerous, disease-causing bacterium from poorly cooked meat, most commonly hamburger, and the reason the disease is often called the hamburger disease. *E. coli* causes bloody diarrhea and cramps and blood and kidney disease in children. The most common cause of the disease is contamination of the meat from intestinal fluids, spilled or smeared on the meat during field dressing or butchering. This

LEFT: Starting with safe meat is a necessity. Be aware of any possible diseases in wild animals and handle field dressing and butchering properly.

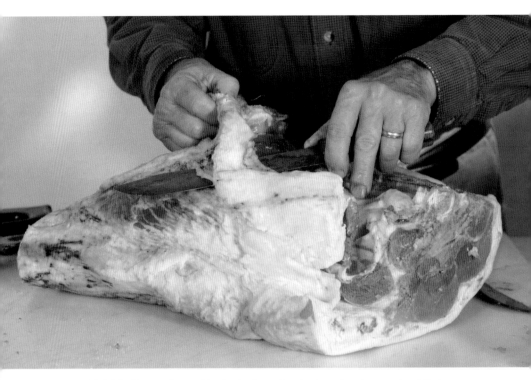

ABOVE: The meat from wild animals, domestic animals, and poultry must be chilled as quickly as possible after killing and before processing and salt curing.

is especially so when the meat is then ground and the contamination is spread throughout. Although commercially butchered meat rarely carries the disease, contamination can happen. There is no reason you can't butcher your own meat as well as field-dress wild game. Just use the proper steps in field dressing and curing for the meat. If you properly process all your own meat, you will know exactly what you and your family are eating.

Regardless of whether you're using meat from wild game or domestic livestock, dress as soon as possible after the animal has been killed to allow the body heat to dissipate rapidly. Wild game, in particular, can be contaminated with fecal bacteria—the degree varying with the hunter's skill and location of the wound, among other factors. Take all the necessary steps to avoid puncturing the digestive tract, a common problem caused by not cutting around and tying off the anus during field dressing, or cutting into the intestines when opening the abdominal cavity. With a gut-shot animal, remove as much digestive material as possible and thoroughly wash out the cavity with lots of running water. Then cut away and discard any meat that has been tainted during the butchering process. Thoroughly clean and disinfect the knife before further use. Do not cut through any organs you suspect to contain disease.

Meat is a very perishable product, with deterioration beginning from the moment the animal is killed and the bleeding is done during butchering. Cool temperatures slow down the process. For this reason, most home butchering must be done in cool or cold weather. It's extremely important to rapidly chill the carcass as soon after killing as possible.

With pork to be salt-cured, this means down to 40°F, as warm meat can spoil before the salt-curing process penetrates to its center. It takes from twelve to fifteen hours to chill a 150-pound hog's carcass down to 40°F, with refrigeration or ambient air temperatures of 32°F to 35°F. Continued exposure to freezing weather, however, causes other problems with uneven chilling. Regardless of variety, if the animal is freshly slaughtered, make sure the meat is chilled rapidly and kept well chilled until it can be salt-cured, cold- or hot-smoked, or cooked.

Cleaning and Disinfecting

One of the most important facets of all steps of butchering, meat processing, salt curing, and smoking is to keep everything clean and disinfected. This includes work surfaces such as tables, countertops, and cutting boards. Make sure to thoroughly clean and disinfect knives, grinders, stuffers, smokers, grill grates, pans, as well as any tools that come into contact with the meat. Clean all surfaces with extremely hot, soapy water with a little bleach added. Then rinse with clean hot water.

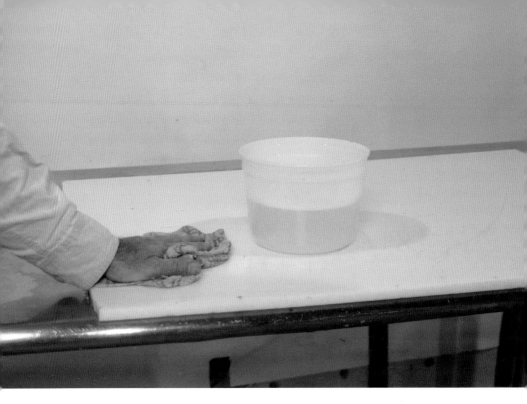

ABOVE: Keep all working surfaces and equipment clean.

BELOW: Disinfect work surfaces and equipment with a mild bleach/water solution and rinse with hot water.

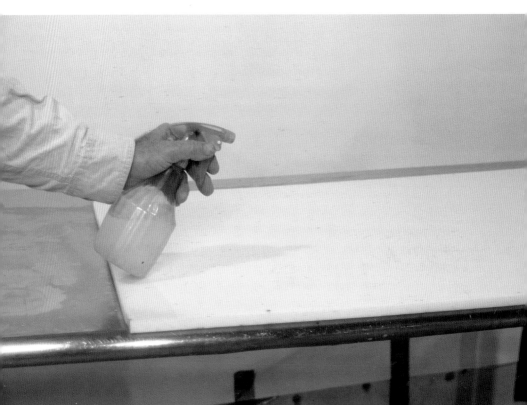

A solution of bleach, soap, and water kept in a spray bottle can also be useful in cleaning surfaces and equipment. Always rinse with clean, hot water afterward. Be sure to clean and sanitize all equipment before using, after using, and before storing away. Make sure your hands and nails are scrupulously clean, or wear food-safe gloves while handling, grinding, mixing, and salting meats.

Choosing Meats

If you're using meat from a home-butchered animal, you have already controlled some of the qualities regarding how the animal, bird, or even fish is raised and then slaughtered. If purchasing meat to salt-cure, cold-smoke, hot-smoke, or cook, there are some qualities to keep in mind.

BELOW: If purchasing meats, choose only top-quality ones. Do not use freezer-burned, or possibly older, meats.

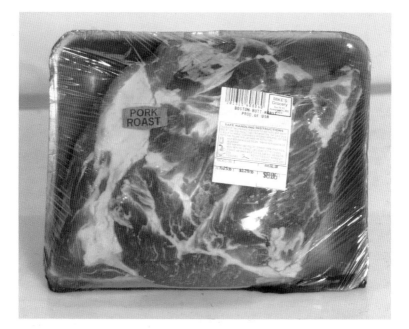

Pork

The quality of pork has changed greatly over the years. In days past, a "fat hog" for butchering often weighed 300 pounds or more, and was just that—a lot of fat. If you look at photos of the ideal hogs from several decades ago, they almost didn't have any legs, with their round, short, barrel-shaped bodies.

These days, hogs have been bred to be extremely lean, with a lot less body fat, long legs, and long frames. Most butcher hogs today are well under 300 pounds. Today's pork tenderloin is just as lean as skinless chicken breast, with 2.98 grams of fat per 3-ounce serving, meeting government guidelines for "extra lean," the reason pork is now called "the other white meat." Quality pork should have a pinkish gray lean meat with streaks of firm white fat. The meat should be fine in texture, and the outer layer of fat should be creamy white and not too thick. The skin should be smooth, free of wrinkles and hair roots. Avoid pork meat that has a coarse texture, is overly red, has white bones, and yellow fat.

Beef

Top-quality beef has a minimum of outer fat, which is creamy white in color. The bones are soft-looking and have a reddish coloration. The meat itself should be firm, fine-textured, and usually a light cherry red. Avoid beef with yellowish or grayish fat, or with heavy marbling (the thicker the marble, the tougher the meat is apt to be). On the other hand, avoid beef with absolutely no marbling and with a deep red color or a two-tone coloration. Avoid beef with a coarse texture and excessive moisture, and beef that is too fresh. (Aging helps develop additional tenderness and flavor.)

Poultry

Choose poultry that is as fresh as possible. Grade A is the best poultry choice and will have nice full bodies with smooth, plump skin. Poultry should not have broken bones, bruises or discoloration, and cuts or excessively bloody areas. In many instances, you'll be purchasing prepackaged poultry products at a supermarket or store. Make sure there are no holes or tears in the packaging. If purchasing frozen poultry, make sure there are no freezer burns and no ice crystals on the package surface. The latter indicates storage may have been less than freezing at some point. Check both the sell-by and use-by dates.

CURING MEATS WITH salt is an age-old skill. The word *sausage* comes from the Latin word *salsus*, which means salted or preserved. Sausage is preserved with salt and was a common food as far back as the Romans. Drying with cold smoke was often done to complete the preservation process as well as add flavor. A number of foods are preserved in this method. For several generations, my family has salted and cured pork hams and bacon in this method, as well as occasionally salt-curing beef. Hams treated in this manner would last a long time without refrigeration, although I have to admit, salty, dry meats are an acquired taste. These days, hams are cured and smoked primarily for

④ The Basics

flavor rather than preservation. Other popular salt-cured meats include bacon, salt pork, jerky, sausages, Canadian bacon, corned beef and corned venison, cured poultry, and any number of salted and smoked wild goose and fish recipes.

Salt Curing

During the curing process, the salt penetrates the meat tissue and draws out the moisture, drying the meat. The drying, as well as the concentration of salt, inhibits the growth of microorganisms. Curing agents, such as nitrate and nitrite, are also usually added.

These days, meats are salt-cured primarily for flavor, rather than for the storage needed before refrigeration.

Although meats can be cured without the curing agents, the best results are with the agents. According to the National Center for Home Food Preservation,

> Nitrates and nitrites are curing agents required to achieve the characteristic flavor, color and stability of cured meat. Nitrate and nitrite are converted to nitric oxide by microorganisms and combine with the meat pigment myoglobin to give the cured meat color.
>
> However, more importantly, nitrite provides protection against the growth of botulism-producing organisms, acts to retard rancidity, and stabilizes the flavor of cured meat. Extreme cautions must be exercised in adding nitrate or nitrite to meat, since too much of either of these ingredients can be toxic to humans. In using these materials, never use more than called for in the recipe. A little is enough. Federal regulations permit a maximum addition of 2.75 ounces of sodium or potassium nitrate per 100 pounds of chopped meat, and 0.25 ounce of sodium or potassium nitrite per 100 pounds of chopped meat. Potassium nitrate (saltpeter) was the salt historically used for curing. However, sodium nitrite

alone, or in combination with nitrate, has largely replaced the straight nitrate cure.

Since these small quantities are difficult to weigh out on most available scales, it is strongly recommended that a commercial premixed cure be used when nitrate or nitrite is called for in the recipe.

The premixes have been diluted with salt so that the small quantities which must be added can more easily be weighed. This reduces the possibility of serious error in handling pure nitrate or nitrite. Several premixes are available. Many local grocery stores stock Morton® Tender Quick® Product and other brands of premix cure. Use this premix as the salt in the recipe and it will supply the needed amount of nitrite simply and safely.

Much controversy has surrounded the use of nitrite in recent years. However, this has been settled and all sausage products produced using nitrite have been shown to be free of

BELOW: Nitrites and nitrates are used to help preserve the meats and provide the characteristic coloring. Using premixed cures is the easiest method of ensuring proper amounts of nitrates and nitrites in the mix.

the known carcinogens. Remember, meats processed without nitrite are more susceptible to bacterial spoilage and flavor changes, and probably should be frozen until used.

Nitrates and nitrites, supplied by potassium nitrate (saltpeter) and Prague powder, have traditionally been used to aid in curing meats. You can purchase the curing ingredients separately and make up your own blend following any number of recipes, or you can purchase premixed, ready-to-use cures and seasonings, with "guaranteed" results. The latter are available from a wide range of butcher supply companies, many on the Internet. Morton Salt's curing products are a longtime favorite for home meat curing and have been used by my family for many years. The Morton Salt family of curing products includes Morton Tender Quick mix, a fast-cure product that has been developed as a cure for meat, poultry, game, salmon, shad, and sablefish. It is a combination of high-grade salt and other quality curing ingredients that can be used for both dry and sweet-pickle curing. Morton Tender Quick mix contains salt, the main preserving agent; sugar; both sodium nitrate and sodium nitrite (curing agents that also contribute to the development of color and flavor); and propylene glycol, to keep the mixture uniform. It can be used interchangeably with Morton Sugar Cure (Plain) mix. It is not a meat tenderizer. Morton Sugar Cure (Plain) mix is formulated for dry or sweet-pickle curing of meat, poultry, game, salmon, shad, and sablefish. It contains salt, propylene glycol, sodium nitrate and sodium nitrite, a blend of natural spices, and dextrose (corn sugar). The *Morton Salt Home Meat Curing Guide* advises,

> Caution: These curing salts are designed to be used at the rate specified in the formulation or recipe. They should not be used at higher levels as results will be inconsistent, cured meats too salty, and the finished products may be unsatisfactory. The curing salts should only be used in meat, poultry, game, salmon, shad and sablefish. Curing salts should not

be substituted for regular salt in other food recipes. Always keep meat refrigerated (36° to 40°F) while curing.

The spices used in Morton Sugar Cure (Plain) mix are packaged separately from the other ingredients. This is to prevent any chemical change that may occur when certain spices and the curing agents are in contact with each other for an extended period of time. If you do not need an entire package of Morton Sugar Cure (Plain) mix for a particular recipe or must make more than one application, prepare a smaller amount by blending 1 1/4 teaspoons of the accompanying spice mix with 1 cup of Morton Sugar Cure (Plain) mix. If any portion of the complete mix with spice is not used within a few days, it should be discarded. It is not necessary to mix the spices with the cure mix if spices are not desired. The Morton Sugar Cure (Plain) mix contains the curing agents and may be used alone. Morton Smoke Flavored Sugar Cure mix is formulated only for dry-curing large cuts of meat, such as hams or bacon. It contains salt, sugar, sodium nitrate, propylene glycol, caramel color, natural hickory smoke flavor, a blend of natural spices, and dextrose (corn sugar). The cure reaction takes longer with this mix than with the Morton Sugar Cure (Plain) mix, so the smoke-flavored product should be used only for dry curing, and is not for making a brine (pickle) solution. The Morton Sausage & Meat Loaf Seasoning Mix is not a curing salt. It is a blend of spices and salt that imparts a delicious flavor to a number of foods. The seasoning mix can be added to sausage, poultry dressing, meat loaf, and casserole dishes, or it can be rubbed on pork, beef, lamb, and poultry before cooking.

Other cure and seasoning mixes are also available. The Sausage Maker carries Maple Ham or Bacon Cure, Honey Ham or Bacon Cure, Country Brown Sugar Cure, as well as Corned Beef Cure. A complete meat-curing kit includes netting, bacon hanger, brine tester, baby dial thermometer, meat pump, stockinette hooks and Honey Ham Bacon Cure, Maple Ham or Bacon Cure, Brown Sugar Cure, Insta Cure No. 1, and a 5-gallon brining bucket.

LEM Products carries a number of cures for pork as well as venison, including Bacon Cure, dry rub; Venison Bacon Cure, dry rub;

Bacon Cure, smoked wet brine; and Venison Bacon Cure, smoked wet brine. Also available from the company are their Backwoods Ham Curing kit and injector. The kits come with Sweeter than Sweet Ham Cure or Brown Sugar Ham kit, which includes two nets for hanging the ham, meat injector for injecting the ham, and complete instructions.

Bradley Smoker has Demerara Cure, Honey Cure, Maple Cure, and Sugar Cure. Hi Mountain Seasonings has their Alaskan Salmon Brine, Buckboard Bacon Cure, Game Bird or Poultry Brine Mix, Gourmet Fish Brine, and Wild River Trout Brine. From Sausage Source come the Luhr-Jensen Brine Mixes.

You can, of course, make up your own cure mixes. Use only pickling or canning salt when making up a cure, and never use iodized table salt. Salt used alone produces a harsh, sometimes bitter, taste. For this reason, sugar, as well as spices, is added for flavor. A wide range of spices can be used for the various cured meats. Most of these are available at your local grocery, but some may need to be purchased from butcher-supply houses. Fresh spices are the best, as they add the most taste.

Many spices lose their flavor over time, especially if kept longer than six months at room temperature. Best storage for spices and

BELOW: A number of premixed cures are available almost anywhere.

ABOVE: You can, of course, make up your own brine cures. Use only pickling or canning salt.

seasonings is below 55°F in airtight containers. Grinding your own fresh spices is the best method for getting the freshest flavor.

General Curing Methods

Meats may be cured using several methods, as well as a combination of methods. These methods include dry curing; pickle, liquid, or immersion curing; and pump curing. Each has its advantages and disadvantages. The old, traditional method of home curing is dry curing, or rubbing the salt and curing agents onto the surface of the meat. This is the method my family has used for generations to cure hams and bacon. A dry sugar cure is the easiest to do—less chance of spoilage, can be done in a wider range of temperatures, and the product

ABOVE: The traditional method of salt curing is with a dry rub.

is generally less perishable. There is more flavor, but also more of a harsh, salty taste. Pickle, or immersion curing, utilizes a salt cure dissolved in water, and the meat is immersed in liquid.

This method is slower; the solution has to be changed every seven days to prevent spoilage, and a salimeter must be used to ensure the salt strength or salinity of the cure pickle is correct. This method has generally been replaced by pump pickling, or pumping the pickle into the meat. The best method for larger cuts such as hams is to use a combination of pumping the hams with the pickle solution and rubbing on the dry cure.

The traditional recipe for dry curing is an 8-3-3 mixture—using 8 pounds of salt, 3 pounds of sugar, and 3 ounces of saltpeter. Another time-tested, less-salty recipe is 5 pounds of salt, 3 pounds of brown or white sugar, 2 3/4 ounces of sodium nitrate, and 1/4 ounce of sodium nitrite (or a total of 3 ounces of nitrate available). Again, do not use more nitrite, as it is toxic. Use 1 ounce of the cure per 1 pound of meat. In our family recipe, we never used saltpeter, nitrate, or nitrites; and the ham was always a gray color. Our hams were traditionally cured, then aged, country-style; and black pepper was also used to help prevent insect infestations during the aging process.

An old-time pickle solution consisted of 8 pounds of salt, 2 pounds of sugar, and 2 ounces of saltpeter dissolved in 4 1/2 gallons of water. The pickle solution was used to immerse the meat.

BELOW: In some cases a pickle cure is used.

Cold Smoking Techniques

Smoking, often combined with salt curing to dry and preserve meats, is one of mankind's oldest and most important skills. In addition, cooking with smoke has also been, and still is, very popular today. As I have said in Chapter 1, the smoke-cooking methods today can either be olden-days primitive or modern-day sophisticated.

First, you must understand the process of smoking. Several types of smoking can be used, depending mostly on the temperature of the heat applied. These include dry or cold smoking and two types of hot smoking, indirect heat and direct heat. Dry or cold smoking is used mostly to help dry out the food and also to impart the flavor of the smoke into the food. This is primarily used for drying jerky and sausages, as well as for drying and adding flavor to hams, bacon, and sometimes fish. Any number of methods and smokers can be used to cold-smoke the meats. In cold smoking, the majority of the heat is kept away from the meat; only the smoke is used. In days past, wooden

BELOW: Salt-cured meats are often cold-smoked for further flavor, to add color, and to aid preservation.

smokehouses, such as the one on my granddad's farm, were used to create the cold smoke used for hams and bacon. Some of today's purchased smokers can also create the needed cold smoke, or you can construct your own backyard cold smoker. The Bradley smoker, with its cold-smoke accessory, is an ideal home cold smoker. Cold smoking is typically a long affair, requiring lots of attention to the chore; but with newer types of smokers, the time is greatly shortened.

There are so many variables in cold smoking that it is not really an exact science. Each fish, fowl, or piece of meat is somewhat different in moisture content, as well as the moisture in the air. It's to your advantage to keep good records of what you smoked, how much it weighed, the amount and type of brining used, and the proper amount of smoking for the end result. Smoking can be as simple or complicated as you like. Primitive people merely used a cured animal skin draped over a framework of saplings and a small smoldering fire.

The following method was used by my family for many years to smoke hams and bacon: Soak the cured meat for about a half hour in cold, fresh water. Use a heavy string to make loops for hanging meat in the smokehouse. Hams and shoulders should be tied through the shank. Light pieces, such as bacon, should be reinforced with wooden skewers. After stringing, scrub the meat clean with hot water and a stiff brush. The water should be about 110°F to 125°F. Then hang the meat up and let it dry. If you don't scrub the meat clean and allow it to dry thoroughly, the smoke won't take evenly, and the meat will be streaked in appearance. The meat should hang overnight in a cool, safe place.

Carefully hang the meat in the smokehouse so none of the pieces are touching. Build a fire in the pit using a hardwood, such as oak, hickory, or apple, or even corncobs. Place a thermometer in the smokehouse and allow the smokehouse to heat up to 100°F, or just hot enough to begin to melt the surface fat on the pork. Then open the top ventilator of the smokehouse just a bit to let out moisture. On the second day, close the ventilator and continue to smoke for at least one more day, or until the meat takes on the desired color. You don't need a dense cloud of smoke—a slight haze will work just as well. The most important thing is not to overheat and cook the meat after the initial start-up—remember you're cold smoking the meat.

The best type of fire in the pit is one built in the Indian fashion of sticks radiating out from the center of a wheel. As the sticks burn, you can continue to push the sticks into the center and add more wood. This type of fire will become cooler and lower rather than hotter as it burns. You can also use twigs or green sawdust to dampen a fire and make it smoke better. One of the main problems with cold smoking is inattention to the fire and the ruined meat in the smokehouse. This takes time and attention. You also need cool to cold ambient temperatures. Smoke only until you achieve the desired nut-brown coloration. Some old-timers in the Ozarks butchered hogs in the early winter, cured and aged the meat until early spring, and then smoked the hams. The bacon was usually already consumed by the time the hams were ready to be smoked.

Using these old ways doesn't always guarantee results. At the end of the smoking period, run a clean, stiff wire into the meat. It should be run in along the bone to the center of the ham and then withdrawn. If it has a sour odor, the meat is tainted. If it doesn't have a sour odor and smells "sweet," the meat is sound. If you get a tainted odor, cut open the meat and examine it for spoilage. If the meat gives off a definite odor of putrefaction, you'll have to discard the entire piece. About the only time I remember my grandfather in a sweat, no pun intended, was when he was waiting for the meat to come out of the smokehouse. In those days of no refrigeration, the smoked meat was really valuable. These days, however, cold smoking of meats is normally done for much shorter times, under more controlled circumstances, and only to add flavor to the meats.

Cold smoking is usually done at 80°F to 100°F. Cold-smoked foods must be further cooked or heat-treated in some means to kill pathogens, the temperature depending on the meat. This may be by conventional oven or by using hot-smoking or smoke-cooking methods. Part III covers hot smoking or smoke cooking of all types of foods, including cold-smoked foods.

EARLY SETTLERS TRADITIONALLY relied on pork for a great deal of their domestic meat. Pigs were economical, fast to grow, and easy to feed. In many instances, the hogs were simply turned out in the woods to fatten on acorn mast. Pigs are also very prolific, providing a lot of meat in a hurry. With an abundance of meat, a means of preservation for extended use became necessary. During the early times, and even today, pork is one of the most popular meats for salt curing and smoking. These pork meats include hams, picnic hams, bacon, Canadian bacon, jowls, and, a necessity for early seafarers as

⑤
Pork

well as trappers and explorers, simple salt pork. Pork sausage is another common salt-cured meat.

Hog butchering days were a tradition in our family, with the entire community coming in for the chore. A dozen or more hogs might be butchered during the day by several families. I remember butchering day as great fun, with lots of excitement. Even the adults seemed to have fun, despite the work. Curing your own pork can still be fun, and a great way of preserving and enjoying a traditional American food.

Curing Ham

Ham is one of the most popular pork cuts for curing. Curing a ham takes some skill, but it is easily learned. Hams can be cured with a homemade recipe or using a purchased cure. The latter contains the proper amounts of nitrites and nitrates, and in some cases, the seasonings as well, making the job easier and the results more consistent. A number of these commercial cures are easily available.

Hams are available in several different types, including fresh or noncured; regular cured; aged, or country-style; and picnic. Hams may or may not be smoked, and they can be cured with different seasonings for a variety of tastes. The first step is to decide what type of ham you wish to cure. How you cut or trim the meat is the first determining factor. The two most common types are the short-cut (regular) ham and the long-cut (country-style) ham.

A short-cut ham is the type most commonly found in grocery stores. It has not been aged and may or may not be smoked. Less salt is used, and the curing time is shorter, creating a milder-tasting product. A long-cut ham is used to make the aged or country-style ham, sometimes called a Virginia or Smithfield ham. A picnic ham is made from the front shoulder of the hog, which is usually smaller. Boneless hams are also very popular, and you can create them as well.

To create a short-cut ham, make the cut separating the ham from the pork side, between the second and third sacral vertebrae, or halfway between the pelvic arch and the end of the pelvic bone. This

BELOW: Hams are cut in two methods, depending on the curing methods, short cut (regular) and long cut (for aged hams).

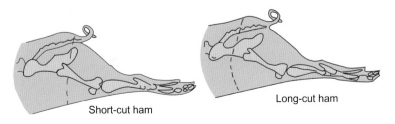

Short-cut ham

Long-cut ham

exposes an open end of the shank bone. Trim the ham by cutting off the first five to six inches of skin from the butt end to about half the distance to the hock end. Cut away excess fat, leaving about one-half inch thick at the butt end.

A long-cut ham to be aged is cut perpendicular to the side and at the pelvic arch, or the bend in the back. This keeps the shank bone intact—with no sponge bone, or marrow, exposed—and provides greater protection from bone souring and from bacteria entering the ham during the aging process. Remove the tail, but leave the skin attached to protect from insects during aging. Excess fat should be removed for proper cure penetration. There will be a slight protrusion of the aitchbone, and this should be sawn off so curing agents can be worked well around the butt end of the ham. Trim up rough corners to smooth and round the ham.

BELOW: Remove the hock.

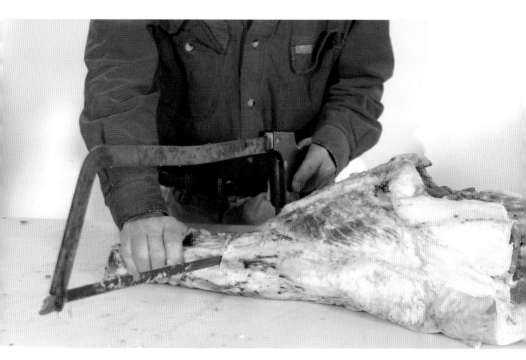

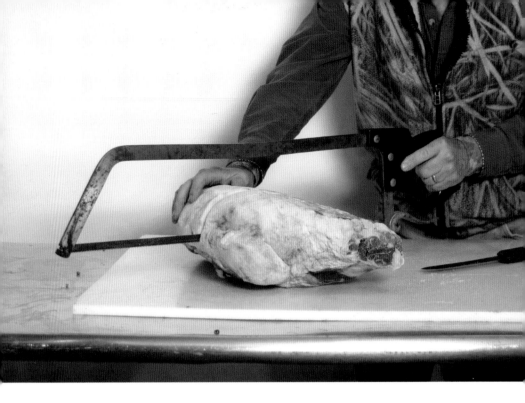

ABOVE: Cut to create the size of ham desired.

BELOW: Trim away excess fat and smooth up the ham.

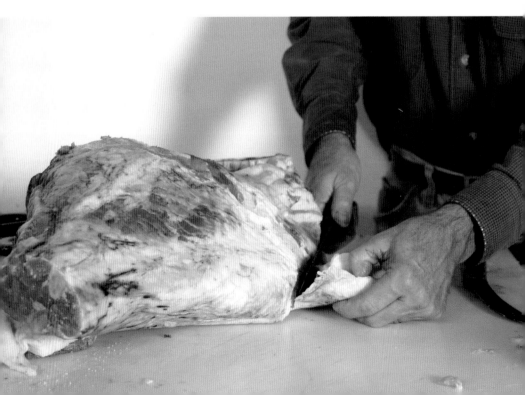

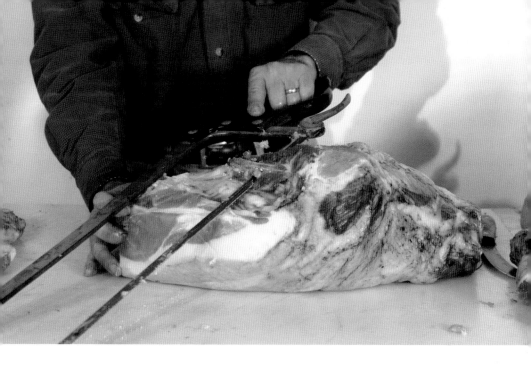

ABOVE: Trim the aitchbone for long-cut hams.

BELOW: The resulting long-cut ham is ready for curing into an aged ham.

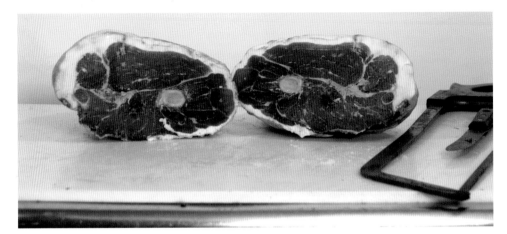

ABOVE: Hams can also be short cut for regular curing.

You can also make a boneless ham, and this is the best tactic for hams weighing more than 25 pounds. Position the ham skin-side down and with the butt end facing you. Using a boning knife, remove the meat from around the aitchbone. Disjoint the aitchbone from the straight long leg bone. Cutting through the top of the ham, remove the leg and shank bone intact, turning the boning knife to trim around the bones as you lift them out. With all bones removed, shape the ham into a roll and tie with white butcher's string. Space the ties crosswise about an inch apart. The boneless roll is treated and cured as for a regular ham.

Hams may be cured using three different processes—dry cure, wet or immersion cure, and the combination cure.

Dry Cure

Dry curing is the traditional and most popular method. Weigh the ham and mix your own cure, or obtain the proper amount of prepackaged cure. Dry-cure hams are cured by rubbing the cure over and into the meat, sometimes with only one application. Other recipes, however, require two to three separate applications. Short-cut, picnic, and long-cut hams require different amounts of cure, different numbers of applications, and different curing times between applications. A

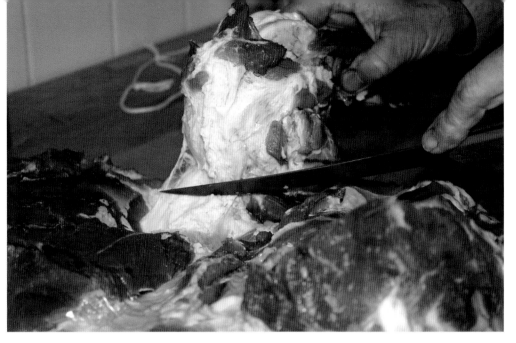

ABOVE: You can also bone out the ham, using a thin-bladed boning knife.

general method is to use three separate rubbings at intervals of three to five days. Picnic hams require only two rubbings, and some aged ham recipes call for only one application.

Divide the cure mixture into two, three, or a single portion, depending on the number of applications. For use with their Tender Quick or Sugar Cure, Morton Salt suggests 3/4 ounce (1 1/2 tablespoons) for each pound of meat. If the ham is long-cut and to be aged,

BELOW: Dry curing is the traditional method of curing a ham. Morton Salt Sugar Cure is used here.

ABOVE: Or you can make up your own cure mix.

use 1 to 1 1/4 ounces (2 to 2 1/2 tablespoons) of curing mix per pound of meat. Measure the cures and have the portions ready to apply before you start working on the ham. Make sure the ham is kept well chilled. Apply the first portion as quickly as the ham can be prepared. Apply the second portion three to five days after the first.

Then for long-cut hams to be aged, make a third application, ten to fourteen days after the first. When rubbing in the cure, make sure all surfaces are well covered, especially the ends. Work the cure in well around the bone ends. Position the ham with the flesh-side up and apply any excess cure on this side.

During the curing process, the ham should be kept at temperatures between 34°F and 40°F. This will usually be in a refrigerator, although some may have access to a cold storage unit. If cold storage space isn't available, you must cure in cold weather. Because of this temperature requirement, we always butchered in late winter in Missouri, as the winter temperature is normally cold and fairly consistent at that time.

To cure, the meat must be placed in a noncorrosive pan or container, or in double-lined plastic food-storage bags. Use only food-storage bags—not nonfood plastic products or garbage bags, as they may contaminate the meat. Position the ham skin-side down on a tray and place in a refrigerator.

The length of curing time for a ham depends on its size as well as the ham type. The common method for dry-curing long-cut hams is 7 days per inch of thickness, or two days per pound. A ham that weighs 16 to 18 pounds and is approximately 5 inches thick at the thickest portion would need to cure for about 35 days. Hams that are short-cut, partially skinned and/or cured using a combination, can be cured at

BELOW: Weigh the ham.

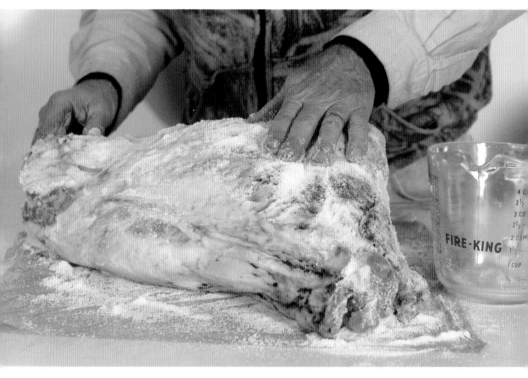

ABOVE: Coat the ham well on all sides with the dry cure mix.

5 days per inch or 1 1/2 days per pound. Don't guess at the cure application dates or curing time. Mark the timing and dates on a calendar posted nearby or on the bag or container.

Immersion Pickle Cure

Immersion or pickle curing is a bit more complicated. The process is slower, and the pickle must be changed every 7 days to prevent spoilage. In the old days, pickling was a common method of curing bits and pieces of pork for salt pork that was used for flavoring and cooking other foods. These days, immersion curing is more precise. A salinity of 75 to 85 percent is suggested, as determined by a salimeter. The percentage must be tested with the water at 60°F, and without the addition of sugar or phosphates. Pickle-curing time for meats immersed in the brine is approximately 3 1/2 to 4 days per pound of ham, or 11 days per inch of thickness, measured at the thickest portion through the center of the ham. For instance, a 15-pound ham would take up to 60 days to cure with the immersion method.

Injection pickle curing speeds up the immersion process with a more complete distribution of the curing liquid into the center of the meat. A portion of the brine is pumped or injected into the meat using two methods: stitch or artery. In the stitch method, a perforated needle is used to inject the cure. In the arterial or artery method, the pickle is pumped directly through the femoral artery of the ham. The pickle should be pumped into the ham at a rate of 10 percent of its weight. A ham weighing 15 pounds requires 1.5 pounds of pickle. Place the meat on a scale during the pumping process to determine when the correct amount has been applied.

BELOW: Place the ham in a large food plastic container or large food bag and place in the refrigerator to cure.

Combination Cure

The best method is the combination cure. This combines pumping the ham with a pickling cure along with rubbing the dry cure mix onto the surface of the meat. With the combination method the cures work from both inside and outside the meat. This makes the meat inside, near the bone, cure more rapidly than with dry-cure alone, thus reducing the chance of spoilage from bone-sour. The remainder of the ham, cured in the traditional dry-cure method, also cures more evenly. And curing time is shortened to about one-third. The combination cure is the preferred method recommended by Morton Salt with their products, and they say, "When used properly, success is almost guaranteed."

The first step is to weigh the ham to determine how much pickle and dry cure mixes to prepare. Prepare a pickle cure by combining one cup of Morton Sugar Cure (Plain) mix or Morton Tender Quick with four cups of water and stir until thoroughly dissolved. Pump the ham with one ounce of pickle cure per pound of meat. If using a Morton Salt pump, four pumps will utilize 16 ounces of pickle cure.

It's important to properly pump the meat. Make sure the meat pump is full to avoid forming air pockets in the meat. Insert the needle to its full length into the meat. Slowly push the pump handle with an even pressure to inject the pickle into the meat near the bone. Slowly and gradually withdraw the needle as you finish the injection, to distribute the cure evenly.

The diagrams of a ham and shoulder show the bone structure. The numbered lines indicate locations where the needle of the meat pump should be inserted into a large ham or a shoulder for five different pumping strokes. If a ham or shoulder is small, eliminate strokes numbered 4 and 5.

Once you have pumped the pickle in place, the Morton Salt dry-cure is applied. The amount of curing mix applied depends on the size of the ham, along with the curing method chosen. "If the hams will be aged, use 3/4 to 1 ounce (1 ½ to 2 tablespoons) of curing mix for each pound of meat. If the ham will not be aged, use ½ ounce (1 tablespoon) per pound of meat." Once you've measured out the proper amount of

ABOVE: Injecting a pickle cure along with the dry cure, or a combination cure, is the best method of curing a ham.

cure, divide it in half and apply half to the ham surface. Five to seven days later, apply the other half of the cure to the ham surface. This allows the cure to penetrate the meat more evenly.

Spices are not required for a successful cure, as the curing agents, sodium nitrate and sodium nitrite, are included in the Morton Salt cure. Spices are included in a separate packet and can be used, mixing them into the cure mix just before application.

Using half of the measured cure mix to be applied, sprinkle the mix on the skin side of the ham and rub it in thoroughly. Turn the ham over, apply the other portion to the meat side, and rub it in well. Make sure to work plenty of the cure mix into both the shank and butt ends and work around any exposed bones. Scoop up and pile any of the surplus cure onto the meat side of the ham.

Again, the combination-cured ham should be cured for about two-thirds of the time needed for a dry-cured ham. After the initial cure, the ham must undergo an equalization period, regardless of whether a dry or a combination cure was used. This allows the salt to further penetrate and spread more evenly throughout the meat to help preserve the ham. If the ham is to be aged, the entire ham should have a salt content of at least 4 percent. If this is achieved, the ham will not spoil or sour, even if temperatures reach as high as 100°F.

For the equalization period, the cured ham should be placed in a tub of clean, cool water and allowed to soak for an hour. This helps to dissolve the curing mix on the surface, distributing the seasoning more evenly. Lightly scrub excess cure from the surface and pat the ham dry. Place the ham into a large clean plastic food bag and return it to the refrigerator for the equalization period. For combination cures, this will take 14 to 15 days; for dry-cured hams, 20 to 25 days. Because the curing agents or salts have been removed from the ham's surface, bacteria may grow there, causing it to become slimy. To help prevent this, keep the bag partially open. This slime growth, however, is not harmful and can be washed or scraped off after the equalization period. A bit of vinegar helps to wash off the mold. After this period, the ham should have shrunk approximately 8 to 10 percent. A short-cut, or regular, ham is now ready to cook and eat or wrap and store in the freezer.

Country Ham Cure

The traditional country, Virginia- or Smithfield-type, aged hams were a staple of the early settlers as these could be stored almost indefinitely. My ancestors were from Virginia, and the hams my grandparents preserved were also aged Virginia-style. Country-style ham has a very different and distinct flavor—very salty, somewhat pungent, and definitely an acquired taste. But the flavor is hard to beat when it's fried for breakfast and served with redeye gravy. A country ham requires additional steps, including aging and smoking if desired.

Some folks like to smoke the ham first and then age, but smoking can destroy some of the enzymes. The flavor and aroma of aged hams come through enzymatic action created by aging the meat in warm weather from 45 to 180 days. Like aging wine or cheese, the flavor comes from the aging process, and is both an art and a science. You'll

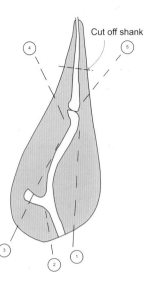

Cut off shank

ABOVE: Using a meat pump, inject the pickle along the bone and in the areas shown.

probably have to experiment a bit until you get the results you desire. The best weather conditions for aging are 75°F to 90°F, and with a relative humidity of 55 to 65 percent. At temperatures over 95°F, the flavor may not develop properly, which is another reason many country people butchered in December. Not only is the cold weather required to properly chill the meat during slaughter and then cure it properly before it spoils, but spring also offers the right conditions for aging the hams. Aged properly, the ham may lose 8 to 12 percent of initial weight during the process. Before aging, many experts like to recoat the hams to create special flavors. Coatings may include black pepper, molasses, brown sugar, and/or cayenne pepper. The hams should be wrapped with paper, such as plain brown paper, then placed in muslin sacks and hung in a cool, dry area with lots of good air circulation. A drafty old barn used to be a popular location. Do not use plastic bags, plastic wrap, or waxed paper that may not allow for air circulation, and do not age in an airtight room.

Another method is also popular for country-curing hams, and is the method used by our family for generations. We basically eliminate the equalization period, wrapping and hanging the hams to cure and

ABOVE: A traditional country-cure ham is typically cured with a dry rub.

age naturally in the Missouri winter, spring, and summer weathers. Another reason is that in the old days, we didn't have the space to cure in the refrigerator. Hams should not, however, be placed outside to cure after January, or in warmer climates. Hams need to cure at least 35 to 45 days with a temperature less than 40°F to prevent spoilage. If freezing occurs during the curing time, allow one additional day for each day/night the temperature drops below freezing.

Burch Family Sugar Cure Recipe

2 cups of salt	4 tbsps. black pepper
1 cup brown sugar	2 tbsps. red or cayenne pepper

Apply at the rate of 1 1/4 ounces per pound of ham.

Lay out brown paper, paper bags or brown wrapping paper, on a clean flat surface. (We have used newspapers for years, but do not use paper with colored inks.) Place the ham on the paper skin-side up and spread the cure mix over the skin. Force some of the cure into the open end of the cut hock. Make sure all surfaces are well covered.

Gently turn the ham over, being careful not to knock off the cure. Coat the cut or flesh side well with the remaining cure mix. Make sure to work the cure well around the cut side and shank end. Work the mix well into the area around the end of the shank bone.

BELOW: Place the ham on clean brown wrapping paper or spread out paper bags or black-and-white newspaper. Cover the ham on all sides with the dry cure, making sure the cure is rubbed well into the joints.

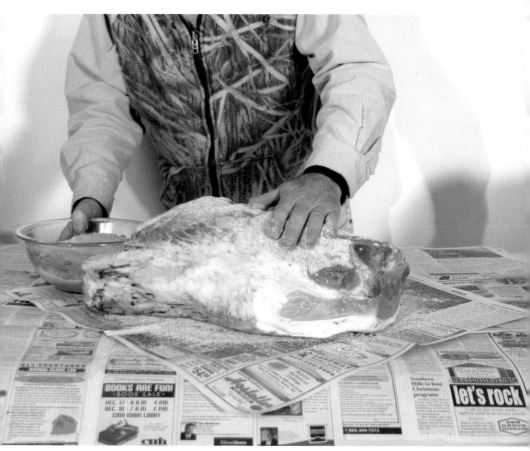

Carefully wrap the paper up around the ham and, holding the paper in place, slide a rectangular homemade muslin bag or purchased stockinette over the paper-wrapped ham. Place the ham in the bag with the shank pointing down to one corner of the bag. Pull the bag tight around the ham and use a stainless steel safety pin to hold it in place. Tie a knot in the bag or tie a heavy cord around the end of the bag.

Allow the bagged ham to sit in place on the table, if it is safe from varmints, for a day to allow the cure to become wetted by the ham liquids. This helps to hold the cure in place.

Hang the ham in a well-ventilated, dry area, again with the shank down, to allow for drainage. The ham will drain, so don't place it over anything that will be damaged by the salty liquid. Do not hang in a moist area with little ventilation, such as a basement or root cellar. If you begin in December, the ham should be cured around the first of April.

You can clean and process the ham for equalization. Then the aging process begins, and the hams can be aged from three to six months. If the hams are in a well-ventilated area, they can also simply be left wrapped from the cure throughout the aging period. This is the method our family has been using. We hang our hams from rafters in an old smokehouse or the barn. We make sure they are well wrapped, and we've never had any insect infestation. It is a good idea to check the hams once a month to make sure rodents or insects have not chewed their way into the hams. Unprotected or stored improperly, ham may be damaged by a number of insect pests, including blowflies, ham mites, cheese or ham skippers, as well as a number of ham beetles.

Smoking

Smoking the ham is not necessary for preservation. Smoking, however, adds to its dark-mahogany appearance, as well as its flavor and preservation. As stated before, you can smoke after curing and before aging, or after aging. Some Smithfield-style hams are smoked after curing, but before aging. Regardless, remove the ham from the bag

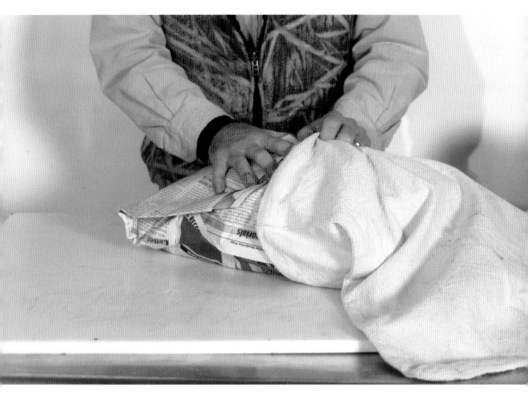

ABOVE: Wrap the paper up around the ham and stuff into a muslin bag.

and wash off all mold growth using a stiff brush and cold water. Pat the ham dry and allow it to hang in the smokehouse or smoker for an hour or two to dry thoroughly. The hams must be smoked with a cold smoke not exceeding 85°F to 100°F. Use only hardwood logs, chips, or sawdust to create a smoldering fire. If smoking more than one ham, make sure that they do not touch each other. Smoke until the hams achieve the color you desire. This will take from 1 to 3 days.

A properly cured and smoked country ham will last just about forever, but it will continue to get drier and saltier. You can clean and freeze the ham for a tastier result. For the best appearance for the ham, remove any mold growth by rubbing it with a cloth dampened with vinegar. Then wash with cold water. Lightly coat the entire ham with vegetable oil and dust with some paprika for additional color.

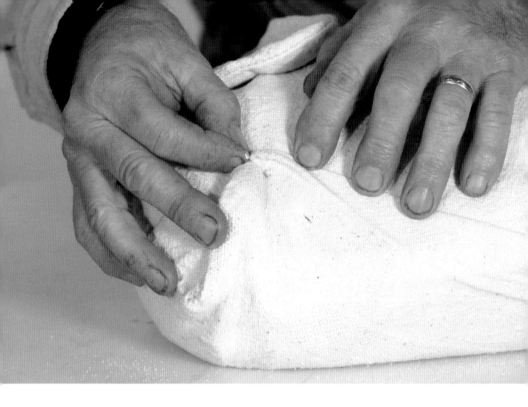

ABOVE: Pull the bottom corners of the bag snug up against the ham and fasten with pins.

BELOW: Tie a heavy cord around the end of the muslin bag.

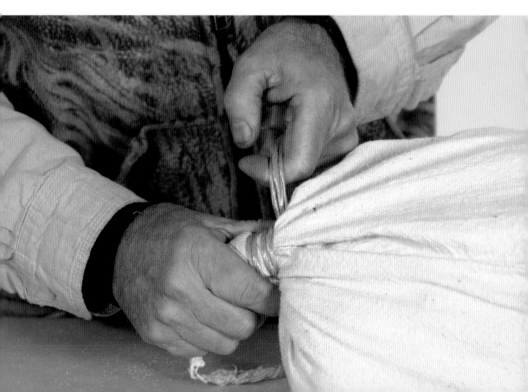

LEFT: Hang the ham in a dry, safe place to age.

Cooking

It takes quite a bit of time to prepare and cure a country ham, and it takes quite a bit of time and effort to properly cook a country ham. Before cooking, wash the ham with cold water; remove any residue salt and cure using a stiff-bristled brush. Place the ham in a large container filled with cold water. Allow to soak overnight or up to 24 hours.

Place the ham in a large container with the skin-side up. Cover with fresh, cold water, and then cover the container with a lid and bring the water to a boil.

BELOW: An aged ham is often cold-smoked for additional flavor and preservation. The aged ham must first be unwrapped and cleaned thoroughly before smoking.

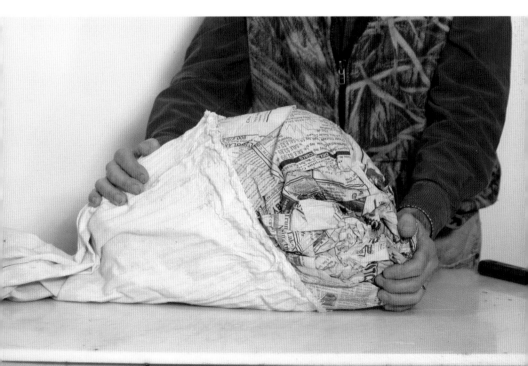

Remove the ham, pour out the water, and repeat the process two to three times. This helps cut down on the salt. Reduce the heat and simmer for 25 minutes per pound, or until the interior temperature reaches 160°F. Remove and allow to drain and cool before carving. Cut away any hardened, dried exterior.

To bake, soak for 24 hours, changing to fresh, cold water as frequently as you can, or at least four times. Trim away hardened meat and skin. Place in a large roaster, fat-side up. Pour two to three inches of water in the roasting pan and bake at 325°F until the internal temperature reaches 160°F, about 25 minutes per pound.

Properly cured and aged, a country ham has a salt content of about 4 percent. You can remove some of the salty taste before cooking by slicing to the desired thickness (1/2 inch for fried ham, 1 inch for ham steaks), and then soaking for several hours in cold water. Change the water frequently to remove as much salt as possible.

Breakfast Ham and Redeye Gravy

Trim the fat from the edges and remove any hardened, dried meat. Cut into the edges about one-quarter inch around each ham slice to prevent the edges from curling. Dice a little ham fat and render it in a hot skillet. Add the ham slices, lower the heat to medium, and cook for about 10 minutes, turning frequently to prevent burning. Remove the ham slices and pour in a little water. Stir frequently and simmer for several minutes to deglaze the pan. Serve with the gravy poured over the ham slices. Old-timers served this with homemade biscuits to sop up the tasty gravy.

Home-Curing Bacon

Bacon is actually made of fresh pork bellies, although they are often called "sides." The cut is made by peeling the spareribs away from the side meat, leaving a portion of it as well as the bellies. Then trim the lean meat in the shoulder area so the lean meat over the entire side is approximately the same thickness.

Remove any areas that do not have lean streaks, such as the bellies, and square up the sides. Be sure to cut the sides to fit your curing equipment. Small sides can be cured in a refrigerator. If curing larger sides, or refrigeration space isn't available, place the sides in the cure from late December through early February to cut the risk of spoilage. If the sides have been scalded and scraped with the skins still on, leave the skins on for the curing process. Sides without skin will cure just as well as those with the skin on. Pork bellies must be chilled to about 40°F within 24 to 30 hours of slaughter.

As a kid, I thought our home-cured bacon was quite salty. Over time, I've experimented with different recipes to achieve a milder cure.

A good mild cure recipe consists of:

7 pounds canning or pickling salt	3 ounces saltpeter, optional (However,
4 pounds light brown sugar	without the saltpeter, the lean meat of
	the bacon won't have the traditional
	red color.)

Apply the cure at the rate of one-half ounce per pound of fresh sides.

In the old days, sides were cured in a cold, dry room, free from pests and the family cat or dog. The sides were covered with a muslin cloth and simply placed on a table that was angled slightly to allow the liquids to drain off.

These days, whether using refrigeration or not, wrapping in food-grade plastic can help deter pests, and is especially helpful when curing in a refrigerator. Weigh the fresh side and prepare the cure. Place the fresh side, skin-side up, on a clean, smooth surface covered with plastic food wrap.

Sprinkle half the cure mixture over the skin side and rub it well into the flesh, coating the edges as well. Turn the side over and sprinkle the remaining cure over the meat side and again coating the ends. Wrap the plastic around the sides and place in a noncorrosive pan or dish. Refrigerate or store in a cold place to cure. If curing outside or

Home-curing bacon is easy and provides a tasty traditional salt-cured meat.

where the sides may freeze during curing, allow the sides to defrost and add one day of curing time for each day they were frozen. Most recipes call for curing the sides 7 days per inch of thickness.

After the bacon has cured for the proper amount of time, remove from the bag, scrub off the salt and cure with lukewarm water, and pat dry with a clean towel. Place back in the refrigerator for a day or two to allow the sides to dry. Cut into 1- to 2-pound packages, then slice to the desired thickness. Wrap and freeze any bacon that won't be consumed within one week. Bacon is also often cold-smoked for flavor.

The Bradley smoker comes with an excellent recipe for home-cured smoked bacon using their Maple Cure. The recipe is for 5 pounds of fresh side or pork belly.

3 tbsps. Bradley Maple Cure (do not use more than this amount for 5 pounds)	1 tsp. ground white pepper
1 tsp. onion granules or onion powder	1 to 3 tbsps. maple syrup (optional)
1 tsp. garlic granules or garlic powder	1/2 to 1 tsp. imitation maple flavor (optional)

Note: If the meat weighs either more or less than 5 pounds, the amount of cure mix applied must be proportional to that weight.

Weigh the pork. If more than one curing container is used, separately calculate the total weight of the meat that will be placed in each container. Refrigerate the meat while the cure mix is being prepared. Any plastic food container with a tight-fitting lid—or a strong plastic bag—can be used as a curing container. Prepare, calculate, and measure the required amount of curing mixture for each container. Mix this curing blend until uniform. Place the meat in the curing container(s). Rub the cure mix on all surfaces evenly. Cover and

BELOW: You can make up your own cure or use a premixed cure such as the Western Legends Buckboard Bacon Cure.

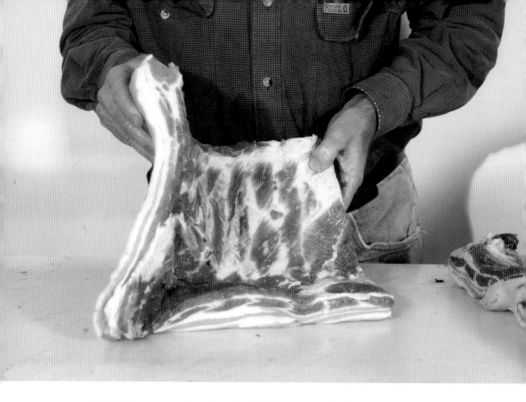

ABOVE: Bacon comes from the side and belly, and excess belly fat must be cut away and the side trimmed up.

BELOW: Regardless of the cure used, the first step is to weigh the side to know how much cure to apply.

ABOVE: Place the side on plastic food wrap and apply the cure.

BELOW: Turn the side over and apply cure to the opposite side.

12/18 ■ Morton Salt
Sugar cure
7 days per inch thickness

ABOVE: Wrap the plastic around the side, place in a plastic food container, and place the container in the refrigerator to cure.

refrigerate. The refrigerator temperature should be set between 34° and 40°F (2.2° to 4.4°C).

Overhaul the pieces of meat after about 12 hours of curing. (*Overhaul* means to rub the surfaces of the meat to redistribute the cure.) Be sure to wet the meat with any liquid that may have accumulated at the bottom of the curing container. Overhaul the meat about every other day until the required curing time has elapsed. Cure one week per inch: If the thickest piece is one inch, cure one week; if the thickest piece is two inches, cure the whole batch two weeks.

When the curing is finished, rinse each piece of pork very well in lukewarm water. Drain in a colander and blot with a clean, dry towel.

ABOVE: After curing, remove from the refrigerator and soak in cold water.

BELOW: Pat the bacon dry with paper towels, then place it back in the refrigerator a day or two so the salt can equalize.

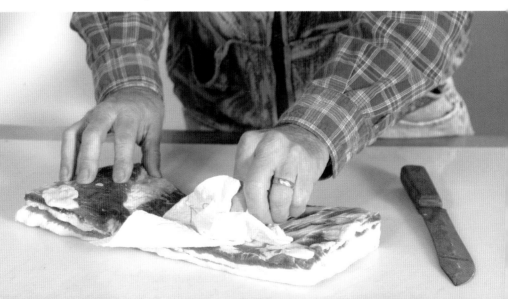

Wrap each piece in a paper towel, and then wrap again with newspaper and refrigerate overnight.

The next morning, remove the paper and dry the surface of the meat in front of an electric fan or inside a smoker heated to about 140°F (60°C). If you're using a smoker, make sure the damper is fully open. Do not use smoke when drying. Drying the surface will require 1 to 2 hours.When the surface is dry, cold-smoke the sides for three hours. If the smoke chamber temperature is higher than 85°F, shorten the smoking time to prevent excessive drying. Raise the smoke chamber temperature to about 150°F. Smoke for two or three hours more or until the surface of the bacon takes on an attractive reddish brown color. Remove the meat and let it cool at room temperature for about one hour. After cooling, place the hunks of bacon in a container—uncovered—and chill overnight. The bacon may be sliced the following morning. Bacon that will not be consumed within about a week should be frozen. The bacon must be cooked before consumption.

Note: If the salt taste is too mild, the next time you make this product, add about 1 teaspoon of cooking or canning salt to the ingredients (do not add more cure). If the salt taste is too strong, reduce the amount of Bradley cure by about 1 teaspoon. This is uncooked meat, and therefore must be cooked before consumption.

Morton Salt Sugar-Cured Bacon

We've been using Morton Salt Sugar Cure (Plain) mix for curing bacon for a long time. It's easy and consistent, and instructions are written on the package. You'll need 1/2 ounce Sugar Cure mix per pound of meat. Place meat on a piece of food plastic large enough to cover both sides. Sprinkle and rub the skin side thoroughly with half the mix, covering the edges of the meat as well. Turn the meat over and place it and the plastic wrap in a plastic food container or in a noncorrosive tray.

Sprinkle the remainder of the cure over the muscle side of the meat. Rub well to distribute and wrap tightly with the plastic wrap. Place in a refrigerator at 36° to 40°F. Allow to cure for 7 days per inch of thickness. Most sides will be from 1 1/2 to 1 3/4 inches thick. On

BELOW: An electric smoker does a great job of cold-smoking bacon.

the fifth day, turn the meat over. At the end of the process, remove the cure, wash the meat in clean, cold water, and place it in the refrigerator for 2 days to equalize. Remove, cut, slice, and freeze.

Hi Mountain Buckboard Bacon Cure

If you don't butcher your own pig, you can sometimes purchase whole pork sides or bellies from a butcher house or your grocery store meat market to make bacon. If you want to try your hand at making bacon but can't find the bellies, Hi Mountain Seasonings suggests using their Buckboard Bacon Cure on a Boston butt pork roast, which is readily available at most supermarkets. This produces an extremely meaty and tasty bacon. You'll need a 4- to 6-pound roast.

First step is to debone and trim the roast. Lay it on a clean cutting surface with the shoulder blade bone to the right. Using a sharp boning knife, remove the shoulder blade bone, keeping the knife as close to the bone as possible. The shoulder blade bone is flat on one side, thus easy to bone out. The opposite side, next to the skin, has a ridge running through the center of it. It takes a bit of patience to cut around this ridge.

With the bone removed, turn the roast fat-side up and trim off any scraps left from boning. Remove the excess fat from the meat. The best

BELOW: Before slicing and freezing, trim the bacon to sizes that will fit your slicer.

ABOVE: Slice and freeze the bacon. The end pieces can be packed as "scrap bacon" and used for dicing and cooking. Vacuum packing helps bacon keep fresher longer in the freezer.

curing thickness is from 3 to 3 1/2 inches. Any excess thickness should be trimmed off the back.

Use 16 ounces of Hi Mountain's Buckboard Bacon Cure for each 25 pounds of meat (1 tablespoon plus 1 1/4 teaspoons per pound). Apply the cure to all surfaces of the meat.

Apply the cure to the meat and massage thoroughly, paying particular attention to the sides and ends, and don't forget the cavity left by boning. Leave excess cure on the meaty side of the roast. Place prepared meat in a nonmetallic container. Cover the container with plastic wrap and place in the refrigerator. The proper temperature is 40°F to 45°F. Let the meat stand in the refrigerator at least 10 days, turning on the fifth day.

After curing, remove the meat from the pan, discard any accumulated liquid, and soak in clean, cool water for 1 to 2 hours. Drain and rinse with fresh water, making sure excess cure is removed (rinse cavity thoroughly). Pat dry and let stand at room temperature at least 1 hour. Shape meat, tucking in the edges. Place on a smoking screen or grill. Insert an internal meat thermometer and place meat in a smoker. Heat smoker to 150 degrees for 45 minutes without any smoke. Increase temperature to 200 degrees and start the smoke. Smoke until the internal temperature of the meat reaches 140 degrees. Turn off heat and leave bacon in smoker for 1 hour to cool down.

All smokehouses are different. These differences play an important role in the smoking and the final flavor. Hi Mountain encourages you to explore the capabilities of your smoker by adjusting times, temperatures, and different smoking chips. In some cases, you may want to use the smoker for flavor only, while achieving the necessary heat by conventional means (kitchen oven).

To cook the Hi Mountain Buckboard Cure bacon, slice thin. Buckboard will cook twice as fast as regular bacon, so be careful not to overcook. Try cutting a thicker slice and enjoy a bacon steak.

Sausage

Fresh sausage is a very common salt-cured pork product, and homemade fresh sausage is hard to beat. The first "test batch" of fresh sausage from butchering days is still one of my favorites. Following is the recipe used by our family for several generations.

Burch Family Fresh
Pork Sausage

---

10 lbs. pork trimmings from butchering

1/4 cup coarse or canning salt

1/8 cup ground black pepper

1/2 to 1 tbsp. crushed red pepper, or to taste

We keep this basic sausage mix fairly lean, without adding fat, and use only basic salt and pepper flavorings, but it does have a bit of a kick. You can also add a bit of ground cayenne pepper instead of, or in addition to, the crushed red pepper. We grow our own cayenne peppers just for this recipe. This sausage has a really rich, meaty taste and is equally good for breakfast or as a sausage burger.

Spread the cut chunks of meat out on a work surface and sprinkle with salt and peppers. Run the seasoned meat through a grinder using a 1/8-inch plate. In the past, we stuffed the sausage into muslin casings and froze it for future use. To use, simply slice the muslin casings

BELOW: Salt, along with other flavorings, is used to cure and flavor sausage.

into 1/2-inch-thick patties. We've also simply made the meat into patties and frozen them with wax paper between each patty. The sausage can be stuffed into sheep casings for breakfast links and can also be cold-smoked for more flavor.

More smoked and fresh sausage recipes are available in *The Complete Guide to Sausage Making* by Monte Burch.

Canadian Bacon

Canadian bacon is considered a delicacy by many, but you can easily make it yourself at home. It is made from boneless pork loins, and these are readily available at supermarkets. To start, trim all fat from the loin. If the sinew membrane is still in place, remove it by using a sharp boning knife to fillet it away from the meat. Morton Salt has a good recipe for Canadian bacon.

1 boneless pork loin	1 tsp. sugar per pound of loin
1 tbsp. Morton Tender Quick or Morton Sugar Cure (Plain) mix per pound of loin	

Mix the cure and sugar together thoroughly and rub onto all surfaces of the loin. Place the loin in a food-grade plastic bag and close. Place on a flat pan and in the refrigerator for 3 to 5 days. Remove, soak the loin in cool water for 30 minutes, and brush off any excess cure. Remove from the water and pat dry. Refrigerate uncovered to dry slightly before cooking.

To cook, cut into 1/8-inch-thick slices. Preheat a skillet and brush it with cooking oil. Fry bacon over low heat, turning to brown evenly. It takes about 8 to 10 minutes.

A Canadian style, and also one popular in Ireland, is peameal bacon. After curing and drying the loins, coat the outside with a mixture of cornmeal and black and red pepper to taste. Cover with plastic wrap and refrigerate. Slice and sprinkle slices with more cornmeal before frying.

ABOVE: You can also cure pork loins to make your own Canadian bacon.

You can also smoke-cook the cured loin to 165°F in a smoker to create a brown-and-serve-style Canadian bacon.

BELOW: Weigh the loins.

ABOVE: A dry cure is applied and the loins placed in a plastic food bag or plastic wrap, then in a food container and cured in the refrigerator.

BELOW: After curing, wash the excess cure from the bacon and slice. Lightly fry to serve, or freeze for use later.

ABOVE: A form of Canadian bacon, peameal bacon is made by coating the Canadian bacon with pepper and cornmeal and then frying it.

Pickled Pigs Feet

The old-timers used every part and parcel of the hog. "Everything but the squeal," were the words Grandmother Burch used to say. Pickled pigs feet or hocks are an old-time food that is a "delicacy" to many. First step is to clean the feet thoroughly, removing the toes and dew-claws. All the dirt and hair must be removed, as well as the glands between the toes. Cure in a pickle cure of two pounds of Morton Salt Tender Quick per gallon of water. Leave the feet in the cure for 7 to 10 days. Make sure the feet are properly chilled at all times. Remove and wash thoroughly. Place the feet in a pan of hot water and simmer slowly until tender. Cook slowly so the skin doesn't split and pull the feet out of shape.

Chill the feet, and then pack them into a jar and cover with hot spiced vinegar. Here's a good recipe:

4 cleaned, cured, and cooked pigs feet	1 tsp. mustard seed
1 tsp. cayenne pepper	1 tsp. coriander seed
1 tsp. onion powder	1/4 tsp. cloves
2 bay leaves	Vinegar to cover
1 tsp. black peppercorns	

Place the feet in a glass jar or crock. Boil the vinegar and spices and pour over the feet. Refrigerate one week. Once opened, the jar must be refrigerated.

BELOW: Pickled pigs feet are made from the feet and hock portions of the leg.

You can also simply cover the pigs feet with cold water, add 2 tablespoons of salt to the recipe, and simmer the feet. Boil the vinegar brine, pour over the cooked feet, and refrigerate.

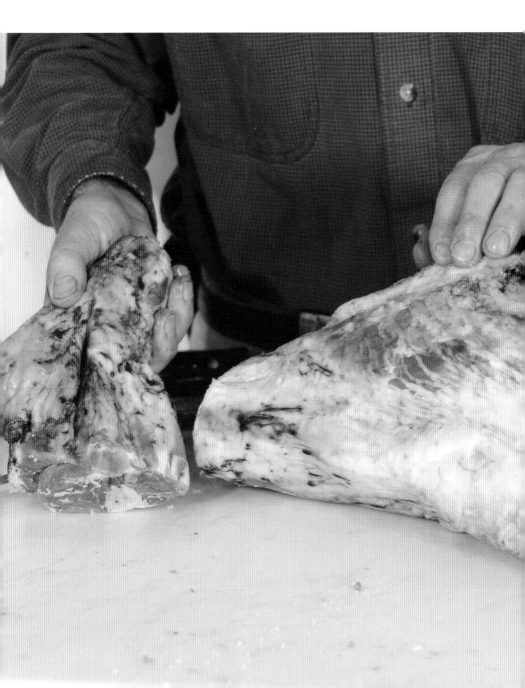

TRADITIONALLY, FEW CUTS of beef were salt-cured or smoked. This was partly due to the cost of producing beef, and because beef tends to be a drier meat. Almost any beef cut can, however, be cured, and even smoked. The most common cuts to cure are the brisket or a rolled plate. Corned beef and pastrami are traditional favorites and beef jerky or salted and dried beef was a staple for many early Americans in the West. Summer sausage is also a popular cured beef product. Cured beef tongue is considered a delicacy. Veal or a beef calf doesn't do as well cured or dried because it doesn't contain as much fat. Following are some traditional beef recipes.

6
Beef

Corned Beef

One of the most popular and common cured beef products is corned beef. This is typically made from the brisket, a tougher meat cut that is also often slow smoke-cooked without curing. Other beef cuts can also be corned, including the rolled plate, chuck, or shoulder. Trim as much fat as possible from the meat and then debone. Beef can be cured using either the immersion or dry-rub method. A number of recipes are available for corning beef, but the following are traditional.

Traditional Corned Beef

25 lbs. of beef

3 lbs. pickling or canning salt

2 cups sugar

1 tbsp. baking soda

This old-time method was often used with trimmings and smaller cuts of beef from slaughter and was common in the days before refrigeration. The beef was left in the pickle until used, if used within a reasonable time. The cured meat was also sometimes pressure-canned, or even air-dried, as jerky. Note: This recipe does not include nitrites or nitrates, and the meat will not be the traditional red color.

Spread a layer of pickling or canning salt over the bottom of a crock or barrel. Rub pieces of meat well with the salt and make a single layer of meat in the bottom. Sprinkle more salt over the layer and continue to layer with salted meat pieces and salt. Allow the salted meat to stand for 24 hours, then cover with a solution made of 2 cups sugar and 1 tablespoon of baking soda per gallon of water. Place a plate with a heavy object on top to hold the meat submerged in the solution. Leave meat for 4 to 6 weeks at 38°F. Make sure to check the meat frequently. If the brine becomes ropey—which may happen if the temperature rises above 38°F—remove the meat, wash thoroughly in warm water, and repack meat in a clean new container, or in the cleaned original container. Cover with a new brine. At the end of the brining time, remove the meat, wash, and pressure-can, cook, or freeze. The meat must be cooked before consumption.

Morton Salt Corned Beef

Another traditional corned beef recipe utilizes a curing pickle of Morton Salt Tender Quick. First step is to make a pickle brine of two pounds of Morton Salt Tender Quick to each gallon of water. Boil the water first, then stir the cure into the water until it dissolves. Allow the brine to chill before using.

Pack the meat pieces into a clean crock or other nonmetallic container. Pour in the Morton Salt Tender Quick curing pickle until it covers the meat and the chunks begin to float. Place a clean plate on the meat and a weight on top to hold the plate down in the pickle cure.

Cure the meat 5 to 6 days; then overhaul, or pour off the pickle, and repack the meat. Take out all the meat pieces and rearrange them so the top pieces are on the bottom and the bottom pieces are on the top. Weight the meat back down and pour the pickle solution back over it. The meat should cure in the pickle for about 2 days per pound of individual pieces. For instance, pieces that weigh 6 pounds should cure for 12 days. Smaller, 4-pound pieces will cure in 8 days. After curing, remove the pieces, wash in warm water, and pressure-can, cook to use immediately, or freeze.

Easy-Does-It Pickle Cure Corned Beef

You can also create a corned beef quite easily in your refrigerator using a 3- to 5-pound beef brisket purchased at your local supermarket. The brisket should be of good quality and well trimmed with no excess fat. Large briskets should be cut in half for easier storing and curing. The brisket is cured using a pickle cure.

BELOW: Corned beef is a traditional and popular salt-cured meat. And corning beef is quite easy to do.

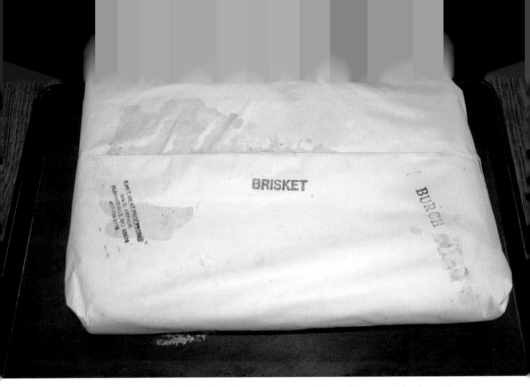

ABOVE: A trimmed beef brisket is the common cut used for making corned beef as well as pastrami.

BELOW: The brisket is cut into two pieces for easy processing, weighed, and a cure is applied by hand.

ABOVE: The cure-treated brisket is placed in a plastic food bag and cured in the refrigerator.

2 cups Morton Salt Tender Quick	2 tsps. black peppercorns
1 cup brown sugar	1 tsp. mustard seeds
3 quarts water	1 tsp. whole cloves

Boil the water, then reduce to a simmer and stir in the Tender Quick and spices. Allow the brine to cool completely. Place the brisket pieces in a nonmetallic container and cover with the pickle cure. Weight the briskets down. Or place the briskets in large food-container bags, add the pickle, and seal well. In this case, place the bagged briskets into a tray. Each day turn the briskets in their pickle mix. Curing time is 7 to 10 days.

After curing, wash and pressure-can, cook, or freeze.

Deli-Style Corned Beef

The Morton Salt folks also have an even easier, dry-rub method of preparing corned beef.

Remove the beef from the refrigerator and wash off any remaining cure.

4–6 lbs. beef brisket

5 tbsps. Morton Tender Quick or Morton
    Sugar Cure (Plain) mix

2 tbsps. brown sugar

1 tbsp. ground black pepper

1 tsp. ground paprika

1 tsp. ground bay leaves

1 tsp. ground allspice

1/2 tsp. garlic powder

Trim the surface of fat from the brisket. In a small bowl, mix Morton Tender Quick mix or Morton Sugar Cure (Plain) mix, sugar, and spices. Rub the mixture into all sides of the brisket. Place the brisket in a food-grade plastic bag and close the end securely. Refrigerate and cure 5 days per inch of meat thickness at the thickest portion of the brisket.

Place brisket in a Dutch oven or a large pot. Add water to cover. Bring to a boil, reduce heat, and simmer until tender.

Simmer in fresh water, cool, and then thin-slice for serving.

Pastrami

Corned beef is cooked by simmering; pastrami is a beef deli meat that is also flavored by smoke cooking. It also has a spice crust added. The first step is to create the corned beef as above. Again, make sure you use a quality cut of beef brisket.

After the brisket has been corned, rinse in cool water and scrub off any excess cure. Completely submerge the meat in clean, cool water and hold it down with a plate. Leave for about an hour to help remove remaining exterior cure. Make up the following spice crust mix:

1/4 cup cracked black peppercorns	1 tsp. crushed red pepper
1/4 cup crushed coriander	1 tsp. garlic powder
1 tbsp. crushed mustard seed	1 tsp. onion powder

Mix the spice ingredients in a small bowl. Place corned brisket fat-side up on a large piece of plastic food wrap. Apply the crust mix

ABOVE: You can also make your own pastrami for delicious deli sandwiches.

to the brisket and rub over the meat surface and onto the ends and sides as well. Press the ingredients into the meat surface as much as possible. Turn the brisket over and repeat for the opposite side. Wrap the plastic food wrap tightly around the meat and place on a tray. Place in the refrigerator for 24 hours.

BELOW: A brisket is typically used and is first corned.

ABOVE: Then the pastrami crust of spicgu'lu'added.

BELOW: The spiced brisket is placed in a plastic food bag and in the refrigerator for 24 hours.

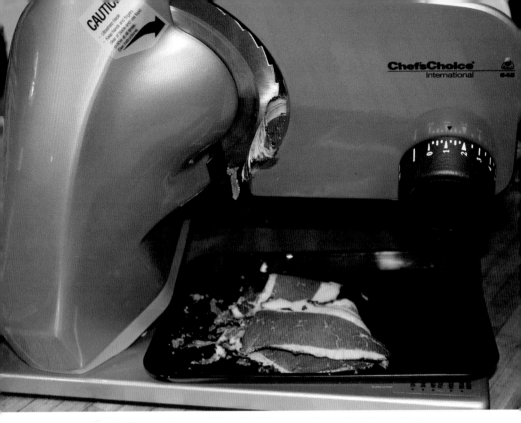

Remove the meat from the refrigerator and pat dry. Place on a smoking tray or screen and place in your smoker. You can also make a small hole in one corner and hang the brisket in the smoker using a smoker hook. Smoke for about an hour at 225°F. Remove and wrap tightly with aluminum foil. Smoke-cook for 2 to 3 hours more or until the internal temperature reaches 160°F. Remove and steam until the meat is tender.

Bradley Smoker Pastrami

You can create a true deli-style smoked pastrami using this recipe from Bradley Smoker. The company suggests using the recipe with either beef or wild game. The recipe utilizes Bradley Sugar Cure mix and is for 5 pounds of meat. Exceedingly fat meat, or exceedingly lean meat, such as beef round, should be avoided.

5 lbs. beef or wild game	1 tsp. oregano
3 tbsps. Bradley Sugar Cure (Do not use more than this amount for 5 pounds.)	1 tsp. paprika
	1/2 tsp. allspice
2 tsps. garlic powder	1/2 tsp. powdered ginger
2 tsps. onion powder	Light corn syrup and coarsely ground
1 tsp. red pepper	black pepper
1 tsp. white pepper	

Note: If the meat weighs more or less than 5 pounds, the amount of cure mix must be proportional to that weight.

Pastrami is thoroughly cooked. It may be steamed, hot-smoked, boiled, or oven-roasted. A modified form of hot water cooking is one of the methods suggested here, but other methods may be employed. Hot smoking or oven roasting can cause excessive drying.

Cut off loose flesh and remove bloody spots and gristle. Remove excess fat. Cut the meat into pieces, the sizes you want to process. The thicker meat requires longer curing time. Rinse all pieces in cold water and drain them in a colander. Blot dry with a paper towel. Measure the thickest chunk of meat and allow 6 days' curing time for every inch of meat. Weigh the meat. If more than one curing container will be used, calculate separately the total weight of the meat that will be placed in each container. Prepare, calculate, and measure the required amount of curing mixture for each container. Evenly rub the cure mix on all meat surfaces. Place the meat in the curing containers. Cover and refrigerate. The refrigerator temperature should be set between 34°F and 40°F. Overhaul the pieces of meat after about 12 hours of curing. Wet the meat with any liquid that may have accumulated at the bottom of the curing container. Overhaul the meat every other day until the required curing time has elapsed. After the curing, rinse each piece of meat in lukewarm water. Drain in a colander and blot with paper towels. Using a basting brush, "paint" each piece of pastrami with light corn syrup or honey diluted with a little water. (This will help the pepper stick to the meat.) Wait a few minutes to allow the surface to become tacky.

Sprinkle and press on coarsely ground pepper. Wrap each piece of beef in a paper towel, and then wrap in newspaper. Refrigerate overnight.

Remove the newspaper and the paper towel and hang the pieces in the smoke chamber, or place on smoking racks. Dry at about 140°F until the surface is dry (about an hour). Do not use smoke during the drying period. To avoid excessive drying and excessively dark coloration, smoke at less than 85 degrees if possible. Smoke the pastrami for 3 to 6 hours, depending on how smoky you want the meat. Raise the temperature to about 145°F for an hour or two toward the end of the smoking time if a darker coloration is desired.

Begin heating water in a large soup pot. Raise the water temperature to 200°F. While the water is heating, wrap each piece of pastrami in plastic food wrap and place in a plastic food cooking bag. Expel as much air as possible from the bag before tying or sealing it. Bag all the meat pieces and put in the hot water cooker at one time. Press and weight them down below the water surface. Maintain the hot water temperature at 200°F and cook the meat about 2 1/2 hours. This long period of cooking will make the meat tender—even gristle will be tender. A thermometer is not required because the cooking time and rather high temperature will ensure the meat is fully cooked. Caution: Raising the water temperature to the boiling point will cause the plastic bags to balloon, the water to overflow the pot, and the meat to shrink excessively. Know that you can maintain the proper water temperature before immersing the bagged meat. If you find it impossible to maintain the correct temperature, simply place the meat pieces in simmering water and slowly simmer until the desired internal temperature is reached.

Remove the meat from the hot water. Carefully open the plastic bags, remove the plastic wrapping, and drain the meat in a colander. Cool at room temperature for about 2 hours and then refrigerate uncovered, overnight. The next morning the pastrami should be sliced, wrapped, and frozen. Freeze the portions that will not be consumed in one week.

Instead of cooking the pastrami in hot water, it can also be roasted in an oven, or steamed. In either case, the pastrami is done when the internal temperature is 170°F. An aluminum foil tent or oven cooking bag should be used if the pastrami is cooked in an oven. If it is steamed, wrap each piece in plastic food wrap before steaming, and use an electronic meat thermometer with a cable probe to monitor the internal temperature. A steamer may be improvised by using a large pan with an elevated rack inside. Cover with a lid.

Corned Tongue

Corned beef tongue was a delicacy in the old days, and it can still provide an unusual delicacy. Actually, both beef and pork tongues were corned, although pork tongues were more often added to sausage or headcheese products, rather than cured. The old-time recipe shown is for a 3- to 4-pound tongue and calls for saltpeter or potassium or sodium nitrate and pickling salt, although you can substitute any commercial curing mix.

1 cup pickling salt	1/4 tsp. ground cloves
1/2 teaspoon saltpeter	4 bay leaves
8 cups water	1/4 cup finely minced onion
2 tbsps. brown sugar	2 cloves garlic, finely minced
1 tbsp. mixed pickling spices	1 medium onion, sliced
1 tsp. paprika	1 tsp. whole black peppercorns
1/2 tsp. freshly ground black pepper	

Create a brine cure by dissolving the salt in the water in a large saucepan. Add in the spices. Boil for 5 minutes and then cool. Place the tongue in a crock or nonmetallic container. Sprinkle the minced garlic and onion over the tongue. Pour the brine solution over the tongue and weight it down with a plate to make sure it is covered and submerged. A water-filled jar can be placed on the plate to keep it down. Place in a refrigerator and allow to cure for 3 weeks at 38°F. Remove and turn the tongue once a week. Remove the cured tongue, rinse off the brine, and place in a large kettle or Dutch oven. Cover with hot water and add the bay leaves, whole peppercorns, and sliced onion. Cover and simmer for 3 to 4 hours (1 hour per pound) or until tender. Remove the tongue and cool. Slit the skin on the underside from the large end to the tip and peel it off. Slice diagonally and chill. Serve cold.

Beef Summer Sausage

Beef was, and still is, salt-cured, and often smoked, to create summer sausage, a delicious snack food. In most instances, pork is added to the mix. This is a great use of trimmings from both beef and pork. One method is to grind the meats and freeze in 1-pound bags. Take out of the freezer, thaw, and make the amount of summer sausage desired. Any number of recipes is available for the various types of summer

sausage. For more recipes, see *The Complete Guide to Sausage Making* by Monte Burch.

3 lbs. beef trimmings	1/4 tsp. ground red pepper
2 lbs. pork trimmings	1 tsp. garlic powder
5 tbsps. curing salt	1 tsp. onion powder
1/2 tsp. ground black pepper	1 cup cold water

Weigh the meats and grind using a 3/16-inch plate. Mix spices and salt and pour over the ground meat. Blend thoroughly. Place in a covered plastic or glass bowl and refrigerate overnight. Roll into logs 1 1/2 x 12 inches long, wrap in plastic or foil or stuff in 2 1/2-inch fibrous casings, and refrigerate overnight again. Cook on oven racks at 325°F, or smoke at 130°F for two hours. Raise temperature to 160 degrees and smoke for 2 more hours. Raise the temperature to 180 degrees and cook until the internal temperature reaches 160 degrees.

LEFT: One of the most popular salt-cured beef products is summer sausage, and it's quick and easy to make.

Beef jerky is another very common and easy-to-make salt-cured beef. The traditional jerky is made by thin-slicing the meat, soaking it in a salt-and-spice mixture, and then drying.

Beef Jerky

A very common use of beef, especially in the West, was to dry it into jerky. This extremely handy food could be carried in saddlebags for a quick meal. These days beef jerky is still a very popular snack food. You can easily make your own beef jerky using the following recipe, which I've been using for almost forty years.

2 lbs. lean beef (round), thinly sliced
1 cup of soy sauce
1 tbsp. garlic powder
1 tbsp. onion powder

1 tsp. black pepper
Ground red pepper, to suit
Water to cover

Although many recipes contain salt as a curing agent, other ingredients can be used to replace the salt. Soy sauce is substituted for the salt in this recipe. Mix the ingredients and place with the meat in a glass bowl, plastic container, or sealable plastic bag and refrigerate for 12 hours or overnight. Remove, drain, and pat dry. Dry in an oven or dehydrator until the meat bends but does not break. To dry in the oven, set on cookie cooling racks over cookie pans, set oven at lowest temperature, and crack door open. For more information on making your own jerky, read *The Complete Jerky Book* by Monte Burch.

ALTHOUGH NOT AS commonly salt-cured and smoked as some other meats, poultry treated in this manner can be extremely delicious. Curing and smoking provides an unusual way to prepare these common and economical meats. Poultry may be smoked, then cooked, cured, and smoke-cooked, or simply smoke-cooked. Poultry may also be processed in salt brine, to which liquid smoke flavoring has been added, and cooked in the oven. This alleviates the smoking process for those without a way of smoking. In all cases, however, poultry must be hot-smoked or smoke-cooked rather than cold-smoked as with some other meats. With poultry, the smoking is for flavor rather than preservation. Poultry that has been cured and smoke-cooked has a unique flavor, has the pink color of smoked foods, and has a slightly increased

Poultry

refrigerator storage life. Cured and smoked poultry can be stored in the refrigerator for up to 2 weeks. Poultry that has not been cured but merely smoke-cooked, however, must be treated and stored just like other cooked poultry. Cured and smoked poultry, like other meats such as ham, can be served hot or cold. The meat also makes delicious sandwiches and salads as well as party snacks.

As with other meats, it's extremely important to start out with quality meat, and that is especially so with poultry, as the meat quickly turns rancid. If the birds are freshly slaughtered, they must be well chilled before curing. All poultry must be chilled to below 40°F immediately after slaughter. Once chilled, the birds must be cured within 2 to 3 days. Purchased processed birds may also be cured and smoked.

ABOVE: Poultry can also be brine-cured and smoked.

In most instances, poultry is brine-cured. This may be by immersion, in the case of smaller poultry, or with the addition of pump curing for turkeys and larger poultry. As it is not a preservative, the salt-brine cure used for poultry is generally milder than for other meats in order to enhance and bring out the natural flavor of the meat.

Like other meats, poultry can be cured using a homemade curing mix or commercially prepared cures. A variety of spices can be used for added seasoning. Some commercial cures also have spices added to the cure, or in separate packets to be added to the cure. Sodium nitrite is usually added to the commercial cures, but can also be added to your own mix. Nitrite provides a light pink coloring after the meat has been cured and heated. Poultry meat that has been smoked without the addition of the nitrite will have more of a tannish white color.

General Brining and Curing Techniques

A number of different curing formulas can be used. If you want to make up your own brine, a basic brine consists of 2 pounds of non-iodized canning or pickling salt, 1 pound of brown sugar, and 3 gallons of water. One tablespoon of liquid smoke can be added if you are unable to smoke the bird but want the smoked flavor.

If you want to use brine with sodium nitrite or saltpeter, add 1.6 ounces of saltpeter to each gallon of water needed to cover the meat. In addition to the basic brine ingredients, seasonings may be added, including black pepper, bay leaves, garlic, basil, oregano, lemon pepper, and thyme.

Use cold water, 36°F to 40°F, to make up the brine. Add ice to cold water to create the volume of chilled water needed to cover the poultry.

BELOW: Poultry is salt-cured only for flavor, and a mild cure is used. You can make up your own, or use a prepared cure.

Add the ingredients, stirring until the solution is clear. Immediately apply the brine to the poultry. Use plastic or noncorrosive containers to hold the brine and poultry. Any poultry can be brined and smoked—whole poultry, boned parts, or parts such as drumsticks and wings. Place the chilled meat in the container, then pour the chilled brine over the meat. Place a plate and a weight over the meat to make sure it is completely submerged. A zippered plastic food bag filled with water and securely closed can also be used as a weight. Make sure the brine goes into the body cavity of whole birds.

In general, the brine will soak into the meat at approximately 1/2 inch per 24 hours at 34°F to 38°F. Small broilers, of 2 pounds or less, as well as small birds such as quail and Cornish hens should be kept in the brine 2 to 3 days. For larger birds, keep in the cure 1 day per pound. Every 3 days, remove the birds from the containers, stir up the brine, and repack the carcasses or pieces back in the brine, shifting the piece around to more evenly distribute the brine.

BELOW: A pickle-cure brine is made up, the poultry submerged in the pickle, and then placed in the refrigerator to cure.

A common brine for large quantities is as follows:

5 lbs. noniodized salt	1 1/8 lbs. cure (containing 6.25 percent
3 lbs. brown sugar	sodium nitrite)
	5 gallons of water

Pack birds into a plastic food-safe or glass container and measure the amount of water needed to cover the birds. Adjust the recipe to suit the amount of water needed to use as cover. Dissolve the salt, sugar, and cure in a portion of the water and then add the rest of the water. A general time to brine a bird is 1 day per pound of individual birds.

A brine for nonsmoked but smoke-flavored chicken is as follows:

1 cup noniodized, canning or pickling salt	1 tsp. oregano
3/4 cup brown sugar	1 tsp. thyme
1 tsp. ground garlic	1 tsp. liquid smoke
1 tsp. dried basil	1 gallon cold water

Soak the poultry in the brine for 24 hours, or 1 day per pound, and then roast in the oven.

Following is a good homemade chicken brine that requires only common pantry items and oven roasting.

3 lemons	5 jalapeño peppers, split and seeded
1 cup canning or pickling salt	1 tbsp. dried basil
3/4 cup brown sugar	1 tbsp. dried oregano
1/4 cup crushed and diced garlic	1 tbsp. dried thyme
3 bay leaves, crushed	1 gallon cold water
1/4 cup whole black peppercorns	

Zest and juice the lemons. Dissolve the salt, sugar, and spices in the water and juice. Brine small fryer chickens for 24 hours, larger birds for 1 day per pound. Keep refrigerated at all times. Rinse, pat dry, and oven-roast.

Injection Brining

Larger birds, such as turkeys, should be cured by the addition of pump, stitch, or injection curing. Otherwise, it takes too long for the cure to work, and results are not as consistent. Weigh the bird and make note of the weight. Stitch-pump the brine into the thickest areas of the breast and thighs, using a large syringe with a no. 12 needle or a meat injector. Make sure you pump the brine into all the muscle areas as well as the joints. To avoid lots of injection holes in the skin, try to use only a few holes for several injections, moving the needle around under the skin as necessary. The carcass should be injected with brine of about 10 to 11 percent of the bird's weight, or until the bird weighs 11 percent more than before injection. Then submerge the carcass into

BELOW: Larger birds, such as turkeys, should be injection-brined as well.

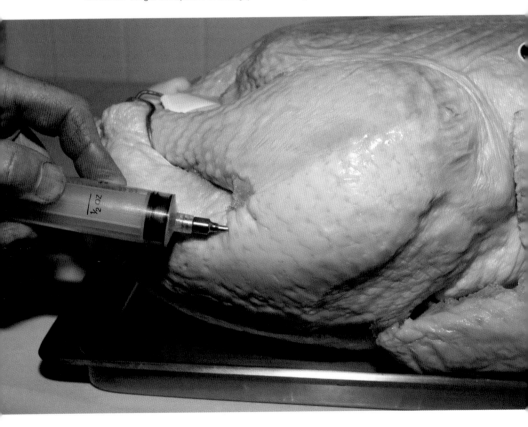

ABOVE: Poultry is rinsed with cold water to remove the brine.

the remaining brine for 2 to 3 days. Make sure the bird is weighted down and kept submerged in the brine. Remove from the brine and wash in cold tap water, allow to soak for an hour or so, changing the water and rewashing frequently. Remove from the water and allow the carcass to dry for about a half hour before cooking.

Smoking and Cooking

A simple method of cooking is to brush liquid smoke over the carcass and then oven-roast at 300°F to 325°F. Cook to an internal temperature of 165°F. Keep covered to retain moisture.

ABOVE: Cured and smoked poultry must be hot-smoked or cooked to an internal temperature of 165°F. A smoked-only turkey must be cooked to an internal temperature of 175°F.

For a true smoke-flavor bird, use the hot-smoke or cook/smoke method with a smoker. Allow the carcass to air-dry thoroughly and place in a stockinette or cheesecloth in order to suspend the bird, or place on the smoker rack. Smoke at 140°F with a very dense smoke for 4 hours. It's also important to maintain a high humidity to prevent the meat from drying excessively. Use dampened wood chips or place a pan of water or marinade over the heat source. Add water as it evaporates. If suspended in the smokehouse, hang tail-end up. After 4 hours, raise the smoker temperature gradually to 190°F to 200°F. After the first 4 hours, the exterior of the carcass should have a nice browned look. To prevent the meat from further drying out, wrap the carcass in aluminum foil. Cook at 190°F to 200°F until the internal temperature of the meat reaches 165 degrees for a cured/smoked turkey, 175°F for a smoke-cooked turkey.

Use a meat thermometer, inserted into the thickest part of the breast, to determine the internal temperature. It will normally take from 12 to 16 hours total to smoke-cook a turkey. You can shorten the time by smoking 4 to 6 hours, or until the outside reaches a nice rich brown color, and then place in a roaster and cook at 300 to 325 degrees in an oven until the internal temperature reaches 165°F for a cured bird and 175°F for a smoke-cooked bird. Cover with foil during oven cooking to prevent the skin from drying out.

Smoke-cooked turkey is extremely perishable and should be treated the same as fresh cooked turkey. Cure-smoked turkey or poultry must be refrigerated, or may be freezer-stored for as long as one year. Use a moisture/vapor-proof bag for storage.

Recipes

Easy-Does-It, Noncured, Smoked Turkey

Bradley Smoker offers some excellent easy-does-it turkey cooking tips with their electric smokers. The Bradley smoker comes with a cold-smoke attachment, and you can use it to add a true-smoked flavor very easily to regular oven-baked turkey. Place the washed, thawed, and dried turkey into the Bradley smoker and place 3 or 4 bisquettes in the smoke unit. Turn on the smoke unit and allow to smoke for 60 to 80 minutes. Once the turkey is smoked, prepare and cook it in your kitchen oven as you normally would. The flavor is amazingly different.

Smoked/Fried Turkey

If you want to add a totally different and delicious flavor to your oil-fried turkey, try cold-smoking it in your Bradley smoker for 60 to 100 minutes before frying. Place the thawed, washed, and dried bird into the smoker and use the cold-smoke unit. After this smoking period, season and deep-fry the bird per your fryer's instructions. You get a deep-fried bird with a rich smoked flavor.

The following recipes are for brining and then both smoking and cooking a 20- to 26-pound turkey in the Bradley smoker. Brining adds flavor and moisture, providing an incredibly rich-flavored and succulent smoked turkey. Your fried-bird-loving friends will beg to learn how you did it.

Brine-Smoked Big Bird

2 cups Bradley Demerara Cure	1 tsp. dried rosemary
1 cup brown sugar	1/2 tsp. dried sage
1/4 cup molasses	2 bay leaves
2 tbsps. whole peppercorns	1 1/2 gallons water
2 tbsps. allspice berries	1/2 gallon apple cider
1 tbsp. whole cloves	1 1/2 cups Jim Beam Bourbon

In a large pot, combine all ingredients and bring to a boil. Then remove the pot and cool to 40°F. Place a clean turkey in a large plastic pail or cooler, breast-side down, and completely cover with the brine. If needed, add more water and apple cider. Place an ice bag on top of the bird to keep it completely submerged, cover the container, and set aside in a refrigerator for 16 hours.

After brining is completed, discard all of the brine solution, dry the bird, and place in a smoker preheated to 205°F to 220°F. You will need to cook the turkey at least 30 minutes per pound or until the internal temperature reaches 165°F. Smoke the bird with Bradley Jim Beam wood bisquettes for about 4 hours at the start of the smoke cooking. It is not recommended to have stuffing in the turkey during the smoke-cooking period.

Brine-Injected Tom Turkey

This recipe is a variation of the previous recipe.

2 cups Bradley Demerara Cure	2 tbsps. minced fresh garlic
2 1/4 cups kosher salt	3 white onions finely chopped
1 1/2 cups brown sugar	9 tbsps. black peppercorns (ground or
3 bunches tarragon	cracked)
3 bunches fresh parsley	6 lemons (halved and squeezed)
6 bay leaves	

Add all the ingredients to 3 gallons of water, heat, and bring to a boil. Stir until salt and sugar are completely dissolved. Chill the brine down to 40°F. Once chilled and ready to use, remove 3 to 6 cups of solution and strain through a paper coffee filter. Inject this mixture into the bird with a food injector. Then place turkey in the remainder of the brine and cover completely using a bag of ice to weigh the bird down. Place in a refrigerator for 16 to 24 hours. Remove the turkey from the brine and discard the brine. Keep refrigerated 12 to 24 hours before smoke cooking. Place the bird into the 200°F to 220°F pre-heated Bradley smoker and smoke for about 4 hours. Then continue cooking for at least 30 minutes per pound or until the internal temperature reaches 150°F. Remove the bird and place it in a conventional oven preheated to 325°F to crisp the skin, and finish cooking to at least 165°F internal temperature.

Cured Roast Chicken

From Morton Salt, this is an easy way to produce a cured and smoke-flavored oven-roasted chicken.

3 lbs. broiler-fryer, whole	2 quarts water
1 cup Morton Tender Quick or Morton Sugar Cure (Plain) mix	Liquid smoke

In a large bowl, dissolve Morton Tender Quick or Morton Sugar Cure (Plain) mix in water. Wash chicken and place in the brine. Weight down chicken with small ceramic plate or bowl so it is completely covered with brine. If necessary, prepare more brine using the same proportions as above. Refrigerate and allow to cure for 24 hours. Rinse chicken thoroughly in cold water to remove excess salt; pat dry. Refrigerate chicken for 6 to 12 hours before cooking so the salt content will equalize.

Place chicken, breast-side up, on a rack in a shallow roasting pan. Brush the chicken with liquid smoke. Roast at 375°F until meat thermometer registers 180°F, about 1 1/4 to 1 3/4 hours, or until the thigh feels soft when pressed between your fingers.

Whole Smoked Chicken

This Bradley Smoker recipe uses any of the Bradley Cures, including Honey, Maple, and Demerara. The recipe is for 5 pounds of chicken. If the meat weighs either more or less than 5 pounds, the amount of cure mix applied must be proportional to the weight, but do not use more than 3 tablespoons per 5 pounds of chicken. For instance, if the weight of the meat is 2 1/2 pounds, then each ingredient, including the Bradley Cure, needs to be cut in half.

The dark meat of the chicken will be pink even when it is fully cooked, and this meat will taste a little like cured ham. You may use any size of bird, or you may mix different sizes of birds. All the birds, regardless of size, may be processed in the same curing container. The sizes are not important because the amount of cure is measured and applied to each bird according to its weight. Use young, tender, well-chilled chickens that are suitable for frying or broiling. Three steps are involved: curing, smoking, and roasting.

5 lbs. chicken	1 tsp. sage, rubbed (packed in the spoon)
3 tbsps. Bradley Cure	1 tsp. oregano
2 tsps. poultry seasoning (packed in the spoon)	1 tsp. ground white pepper
2 tsps. onion powder	1 tsp. paprika
1 tsp. MSG (optional)	1/2 tsp. dill weed
1 tsp. garlic powder	1/2 tsp. crushed bay leaf

Rinse and clean the bird or birds and let drain in a colander. Next, use a sturdy fork to pierce the chicken all over, especially the legs and breast. Prepare the proper amount of cure according to the weight of the bird. (If more than one bird is being cured, prepare the proper amount for each bird.) Apply the cure uniformly to the bird; a shaker with large holes works well for this. Be sure to apply the cure to the inside of the body cavity as well as to the outside skin. Cure the chicken in the refrigerator for at least 4 days. Rub all surfaces to redistribute (overhaul) once a day during that period. The refrigerator temperature should be set between 34°F and 40°F.

At the end of the curing period, rinse the bird inside and out in cool water, and blot it inside and out as well. Stuff the body cavity with paper towels that have been wrapped around crumpled newspapers. Store the bird in the refrigerator overnight, preferably with the tail pointed upward. Put a paper towel and several layers of newspaper under the chicken to absorb the water.

The next morning, set up the smoker to finish drying the chicken. Preheat smoker to about 140°F. If possible, hang the bird in the smoker with the tail pointing up. This allows the melted fat and juices to fall freely into the smoker drip tray instead of collecting in the body cavity. Dry the bird in the smoker at 140°F. After the skin is dry to the touch (about an hour), cold-smoke it for 3 hours at 85°F, or as low a temperature as possible. (The cold-smoke unit is excellent for this.) This will provide a mild smoke flavor. If you like a stronger smoke flavor, smoke the chicken for about 6 hours. Apply cooking oil to the chicken and hot-smoke at 145°F until the bird takes on a beautiful reddish brown color (probably 2 more hours).

Remove the chicken from the smoker. Apply cooking oil to the skin again. Cover well with foil, but do not seal the foil tightly—leave a few openings for steam to escape. (Because the chicken has been browned in the smoker, additional browning is undesirable, and the foil prevents this. The loose wrapping of foil allows some steam to escape, but it also prevents excessive drying.) Add about 1 tablespoon of water to the inside of the foil, and roast the bird in a kitchen oven at 350°F for about 2 hours. Use a meat thermometer to test for doneness. When the internal temperature is 180°F, it is done.

Hi Mountain Game Bird and Poultry Brine Mix

Hi Mountain Seasonings has a very-easy-to-use poultry brine mix. Everything needed, including cure and spices, is included. The brine mix comes in two packets; each will make up 1 gallon of brine. Poultry should be well chilled before curing. Dissolve 1 pouch in 1 gallon of ice water (34°F to 38°F) in a nonmetallic container. Immerse the poultry into the brine, completely covering it. Weight down the bird to keep it submerged. Place in a refrigerator for 24 hours.

Preheat the smoker to 140°F. Remove the poultry from the brine and rinse with fresh tap water. Pat dry and place in a smoker without smoke for the first hour to dry the bird. Add smoke and raise the temperature to 200°F. Smoke until desired internal temperature of 175°F is reached. If you cannot get the poultry to the desired internal temperature in the smoker, when the desired color is reached, place the poultry in the kitchen oven to finish. Smoke time will vary depending on the type of smoker, location, outside temperature, and so forth.

BELOW: Prepared brines are easy to use.

⑧
Fish

SALTED, CURED, PICKLED, and smoked fish are age-old foods that are enjoyed worldwide. Fish preserved using these methods have been a staple of many cultures, including the Native Americans. Even today, catches of many saltwater and larger freshwater fish often result in an excess of fish. Although an excess amount of fresh fish can, of course, be frozen, salting, curing, and smoking are traditional methods of preserving excess amounts of fresh fish. The most important use of curing and smoking these days, however, is more for flavor rather than preservation. It's a great way of producing exquisite gourmet meals such as smoked salmon, trout, tuna, halibut, catfish, and inland stripers, even bass and bluegill, in addition to other seafood delicacies. Smoking fish is also fun and interesting. In earlier days, when curing and smoking was primarily for preservation, large amounts of salt were commonly used. These days salt curing and smoking is mostly for adding flavor and for the sake of appearance. These lightly salted and smoked foods are not, however, preserved, and improper cooking or smoking can cause serious, even fatal, food poisoning. Lightly salted and smoked fish must be kept refrigerated at 38°F, or frozen for future consumption.

Salting and Smoking

Three very important steps must be taken in salting and smoking fish. The first is to make sure the fish is cleaned properly and kept well chilled. Second, fish must be brined long enough to make sure there is an adequate amount of salt in the meat. Third, fish can be cold-smoked and then cooked, or smoke-cooked. Cold smoking, or smoking below 80°F to 90°F, is the most complicated, with long smoking times and less consistent results, with more chances for food safety problems. Most smoked fish these days is smoke-cooked. Fish must be heated to 160°F internal temperature, and this temperature must be maintained at least 30 minutes during smoke cooking or smoking and heating.

Although almost any fish can be smoked, fatty fish can be brined and smoked easier than leaner fish because the lean fish absorb the salt quicker, and it's easy to get the fish too salty and smoked too dry. Good freshwater choices include catfish, suckers, inland stripers, white bass, shad, and sturgeon. Trout are also an excellent choice for smoking. Almost all saltwater species, and especially the more oily species—including mackerel, sablefish, and tuna—can be smoked. Even the smaller fish—such as herring and smelt—can be successfully smoked. The salmon species are some of the most popular and common smoked varieties.

Choosing and Preparing Salmon

If curing and smoking your own freshly caught salmon, do so as quickly as possible, keeping the flesh well chilled or frozen until you are ready to smoke. Clean the salmon and eviscerate. Remove the tail, fins, and head. Cut about 1/2 to 1 inch off the belly on each side. Leave the skin on, or remove as desired. Pull out any pinbones. For small salmon, cut into 3-inch-wide steaks. Then cut the steaks apart along the backbone. For large salmon, cut into 1 1/2- to 2-inch-wide pieces.

If purchasing salmon for smoking, it's important to get good-quality, fresh salmon. The flesh should be evenly colored, with no mottling, and clean and solid. Salmon flesh that is mushy is suspect and will not cure and smoke properly. If the head is still on the fish, the eyes should be bright and clear. The skin should have a bright, shiny appearance. Quite often, however, you will be purchasing prepackaged salmon, and it's more difficult to determine quality. If the salmon is fresh, press on the flesh. If it doesn't spring back when you remove your finger, it's probably spoiled, or "old." The different salmon varieties have different qualities. The salmon with higher oil contents have more flavor and result in a more moist smoked food. These include the sockeye, king, and Atlantic salmon. Salmon with lower oil contents and a lighter flavor include the pink and chum. A great deal of the Atlantic salmon is farm-raised and has a fairly firm, oily flesh.

Smoke only top-quality fish, either fresh or frozen. Freezer-burned or dried-out fish from your freezer won't provide good-quality food. Freshly caught fish should be thoroughly cleaned to remove all blood, slime, and harmful bacteria. Either fillet or split the fish. Either skin or leave the skin on depending on species, recipe, and preference. Larger fish can be cut into steaks. Regardless, the pieces should not be more than 1 inch thick, or they may not smoke properly and then could spoil. Cut all pieces into uniform sizes so all will be salted equally and will smoke evenly. Keep fish to be smoked as cool as possible at all times. The ideal temperature is 38°F. Check your refrigerator to make sure it is at or below this temperature. Avoid cross-contamination by not handling raw fish in the same area where smoked fish is wrapped, prepared, or kept.

Parasites such as tapeworms and nematodes can also be a problem in some fish, especially when you're using low-salt, low-temperature, cold-smoke preserving methods. These parasites can survive some salt brining and smoking methods, creating serious health problems. Freezing the raw fish at a temperature of 0°F for two weeks or longer can destroy the parasites. Check your freezer to make sure it reaches the proper temperature. Frozen fish should be thawed in a cool place or placed in cool water to thaw. Do not refreeze. The best method is to first process the fish, cutting into proper sizes, and then freezing.

Hot smoking or smoke cooking, which means raising the internal temperature to at least 160°F, and holding the fish for 30 minutes at this temperature, kills the parasites as well as other bacteria.

Brining

Fish can be salt-cured by either dry cure or brining. Brining is the most common method and produces the most consistent results. Salt used should be kosher, pickling, and canning, or flake salt. Do not use iodized, rock, or sea salt. These may contain impurities and can cause an off-flavor. Herbs and spices are often added to the brine for flavor. Commercial fish brines are also available, again with a variety of spices and herbs. Cure may or may not be used, depending on the recipe.

A general-purpose brine consists of 1 part salt to 7 parts of water, or 1 cup of salt to 7 cups of water. Make the brine using ice water, or chill the brine to 38°F or lower before using. This will help the flesh take up the salt and reduce bacterial growth. Make sure you have enough

ABOVE: A pickle brine is used, but you can make up your own.

chilled brine to cover the fish. Place the fish in the brine, making sure it is well submerged, and brine for anywhere from 1/2 to 6 hours, or for the time suggested by the recipe.

Equalizing

Remove the fish from the brine and wash thoroughly. Soak for about an hour in fresh, cold water to remove excess brine. The fish must be dried and the salt equalized. This allows the salt to be evenly distributed throughout the fish. The fish should also be allowed to dry

BELOW: Prepared brine mixes are also available.

so the surface forms a tough, shiny coat, or pellicle. This seals in the moisture and provides an evenly smoked appearance without streaks. Keep the fish refrigerated at temperatures below 38°F for salt equalization and to begin the surface-drying process. This may take from 2 to 24 hours, depending on species, recipe, and thickness of the pieces. Normally, 1-inch-thick fillets will take from 6 to 10 hours.

ABOVE: Fish is then hot-smoked. In days past, fish were often cold-smoked, along with salting for preservation. These days, the suggested method is to hot-smoke only to an internal temperature of 160°F.

Smoking

Smoking may be done in water-pan smokers, large barbecue smokers, as well as dedicated wood or electric smokers. Forming the pellicle, or outer coating, usually requires a warmer temperature. This is normally done in the smoker before the smoking begins. A low temperature without smoke is used, and may take from 30 minutes to 2 hours. After the pellicle has formed, the fish is smoked. Two basic methods are used: cold smoke and hot smoke. Hot smoke may be indirect- or direct-heat smoke cooking.

Cold-smoking fish is actually drying out the meat, resulting in an internal temperature of less than 90 degrees. Examples are lox or Nova Scotia–style salmon. Cold smoking results in raw fish, and steps must be taken to kill bacteria and parasites. Cold smoking is difficult in areas with a high humidity and requires long smoking times or several days.

Hot smoking is the most popular form with home fish smokers. Hot smoking may be done with a dedicated smoker, either wood or electric. After the pellicle has formed, the temperature of the smoker is gradually raised to 160°F. Raising temperatures too quickly can cause the fish pieces to break apart. The final smoker temperature should be from 180°F to 225°F, depending on the recipe. The fish should be cooked to an internal temperature of 160°F and held at that temperature for at least 30 minutes, and this should occur within 6 to 8 hours of placing the meat in the smoker. Use a standard meat thermometer or remote thermometer placed in the thickest portion of the fish to test. If the smoker doesn't raise the temperature properly within 8 hours or so, continue cooking the smoked fish in a home oven set at 300°F for final heat treatment. Again, bring the internal temperature to 160°F and bake for at least 30 minutes.

The second method of hot smoking is direct smoke cooking, such as done on a barbecue grill or water-pan smoker. In this case, the smoke-cooking temperatures are usually higher, and smoking times are much shorter.

Storage

Proper storage of smoked fish is also important. Allow the smoked or heated fish to cool at room temperature. Then wrap in plastic wrap and keep refrigerated at below 38°F. The smoked fish can also be pressure-canned, frozen, or, better yet, vacuum-packed and frozen.

Smoked Fish Recipes

Bradley Maple Cured Smoked Salmon

Bradley Smoker is famous for their salmon smoking, and their maple-flavored salmon is excellent served with wild rice and stir-fried vegetables. Or smash and blend with equal amounts of cream cheese for a delicious spread.

1 large salmon fillet	1/4 cup lemon juice
1 quart water	10 whole cloves
1/2 cup pickling and canning salt	10 whole allspice berries
1/2 cup maple syrup	1 bay leaf
1/4 cup dark rum	

In a medium-sized bowl, combine the brine ingredients. Place the salmon fillet in a nonmetallic dish and cover with the brine. Make certain the fish is completely submersed in the brine. Cover and refrigerate for at least 2 hours. Remove salmon from the brine and pat dry with paper towels. Place salmon skin-side down on smoker rack and allow to air-dry for about 1 hour.

Preheat the Bradley smoker to between 180°F and 200°F. Using Bradley maple-flavor bisquettes, smoke-cook the salmon approximately 1 1/2 hours.

Hi Mountain Alaskan Salmon Brine Mix

Another maple-flavored recipe, this utilizes a maple-flavored and spiced brine. The package comes with two pouches of cure. Dissolve 1 pouch in 1 gallon of ice water. Fish should be fresh and chilled before curing. Immerse the fish in the brine solution, making sure it is completely covered. Place in a refrigerator for 24 hours.

Remove fish from the brine, rinse well with fresh, cold water, and pat dry. Let fish sit at room temperature for about 30 minutes, then smoke. Smoking time can vary depending on the type of smoker, location, outside temperature, and so forth. Fish should be smoked until internal temperature reaches 155°F to 160°F.

Bradley's Famous Hot-Smoked Salmon

If you luck into a cooler full of salmon, this recipe is a good choice, and it has been one of Bradley's most popular recipes using their alder bisquettes.

Cure (white sugar and salt—approximately 1 pound of salt to 2 ounces of sugar)	Coarse ground black pepper
Vegetable oil	Dried parsley or chive flakes
Garlic and onion salt or powder (You can substitute dill or dry mustard for the garlic and onion.)	

Leave the skin on the salmon. If fillet is over 1 inch thick, slash the flesh every 2 to 3 inches about 1/2 to 3/8 inch deep, parallel and running in the direction of the rib. Slather fish with a liberal amount of vegetable oil. Sprinkle cure heavily and evenly on the fillet. Use enough cure so that it doesn't wet out in the oil. Sprinkle a moderate amount of desired spices over the fillet. Rub the spices and cure lightly, including any cut surfaces. Sprinkle a moderate amount of coarse black pepper. Wrap two similar-sized salmon fillets, flesh to flesh, with plastic wrap or plastic bag and then place in a cooler or nonmetallic container. Cover fish to ensure air has no access and refrigerate 14 to 20 hours.

Remove fish from the cure and place skin-side down on oiled racks. Rub fillets to even out the residual cure and sprinkle with parsley or chive flakes. Place the racks in the Bradley smoker. Using alder-flavored bisquettes, start the Bradley smoker at a very low temperature (100°F to 120°F) for 1 to 2 hours. After the first 2 hours, increase the temperature to 140° F for 2 to 4 hours. Finish at 175°F to 200°F for 1 to 2 hours.

Smoked Blue Cats

Blue cats can get big, well over 100 pounds, and these big cats can provide large fillets, excellent for smoking. Their flaky, white, but somewhat oily flesh provides a good smoked flavor. The skins should be removed and the fillets trimmed so the meat is an even thickness, cutting away some of the thin belly meat as well as the thinner portion of the tail area. Make up enough brine to cover and submerge the fillets. One gallon of brine will do about 5 pounds of fillets.

1 gallon ice water	1 tsp. garlic powder
1 cup brown sugar	1 tsp. onion powder
1/2 cup canning or pickling salt	1 tsp. ground red pepper
1/2 cup soy sauce	1 tsp. ground black pepper
1/4 cup Worcestershire sauce	

Add the brine ingredients to the gallon of water and heat to dissolve all ingredients. Chill brine and pour over cleaned fillets in a noncorrosive plastic container. Weight the fillets to keep them well submerged. An alternative method is to place the fish and brine in a sealable plastic food bag, and then place this in a container in case the bag leaks. The double-fastening-type bag works best for this tactic. Two-gallon freezer bags will hold the brine and about 5 or 6 pounds of fillets. Place in a refrigerator and leave for about 4 hours at 38°F. Turn the bags or move the fillets around once an hour to make sure the brine is evenly distributed.

Remove from the brine and place on paper towels. Pat dry, then place on smoker racks or other raised racks. Allow to air-dry at room temperature for an hour or so. The surface of the fish should start to become slightly hard, forming a pellicle. Or dry in the smoker with the smoker set at an extremely low temperature.

Gradually increase the temperature to 180°F or 200°F and cook until the fish flakes and the internal temperature reaches 160°F. Remove and cool the fish before serving.

Shake 'n' Smoke Salmon

This is a very simple and easy salmon recipe, but it can also be used with almost any fish. It's a great way of curing and smoking small fillets, such as bluegill.

Simply skin, clean, and prepare the fillets. Pat the fillets dry with paper towels. Place a cup of Morton Salt Sugar Cure mix and a cup of brown sugar in a large plastic food bag. Place the fillets a few at a time into the bag and shake to coat all surfaces. Remove and place in a noncorrosive or plastic bowl and place in the refrigerator for an hour. Remove, rinse in cold water, pat dry, and smoke in your smoker. You can also sprinkle liquid smoke over the cure-covered fillets and simply place them on wire racks in a 325-degree oven until done. Add your favorite seasonings and spices for added flavor.

Smoked Kings

Actually, any salmon can be used with this delicious recipe, but it's great with the big kings, also called chinook. The following brine will cure about 4 to 5 pounds of salmon.

3 quarts cold water	1/2 cup honey
1 quart soy sauce	1 tsp. garlic powder
2 cups brown sugar	1 tsp. onion powder
1 cup pickling or canning salt	1 tbsp. ground white pepper

Mix ingredients in a saucepan, heat and stir until dissolved, and then chill in a refrigerator. Place salmon chunks in the brine in a non-corrosive container in the refrigerator. Salmon less than 1 inch thick will brine in 6 to 8 hours. Salmon 1 inch or more thick will require 12 hours' or overnight brining.

Remove from the brine, rinse, allow to air-dry for a couple of hours, and then hot-smoke to an internal temperature of 160°F.

Lemon Pepper Smoked Striper

The oily-white meat of ocean-run or inland stripers is a great choice for smoking. Large white bass or hybrids also have an oily flesh and can also be smoked. Prepare the fish by filleting and cutting into chunks.

1 gallon water	1 tsp. garlic powder
2 cups brown sugar	1 tsp. onion powder
1/2 cup salt	1/2 tsp. ginger
1/2 cup reconstituted lemon juice	1 tbsp. lemon pepper

Place cleaned and prepared fish chunks in chilled brine overnight in a refrigerator that is cooler than 38°F. Remove from the brine and rinse the fish in cold water. Dry the fish and hot-smoke until the internal temperature reaches 160°F.

Hot-Smoked Bluefish

If you have an abundance of blues, and who doesn't when the bite is on, this is a great recipe for doing 20 or so pounds of fish. Clean the fish and cut into equal-sized fillets, leaving the skins on. Make up the brine to cover fish, using 1/2 pound of salt and 1 pound of brown sugar to each gallon of water needed to cover the fish. Brine overnight in a refrigerator. Remove from the brine, rinse, and pat dry. Place on smoker racks with the skin-side down and hot-smoke until an internal temperature of 160°F is reached. With a temperature-controlled smoker, this should average 4 hours at 120°F, plus 2 hours at 180 degrees.

Smoked Tuna Steaks

Tuna steaks are also great salted and smoked. Cover with the following brine:

1/2 gallon water	1 tsp. reconstituted lemon juice
1/2 cup salt	1 tbsp. ground white pepper
1 cup brown sugar	1/4 tsp. ground ginger
6 bay leaves (crumbled)	

Mix ingredients in water, simmer until dissolved, and then chill the brine. Place tuna steaks in a large plastic sealable food bag, add chilled brine, and place in a refrigerator for 4 hours. Turn once every hour.

Remove from the brine, rinse, pat dry, and allow to air-dry for about a half hour. Brush with vegetable oil and place in smoker. This is an excellent recipe to use with a water smoker. Just add white wine to the water pan. Smoke for 3 or 4 hours, or until tender and flaky.

Smoked Oysters

If you want a real treat, try smoked oysters. This recipe is for 1/2 gallon of oysters. Rinse the oysters and drain them in a colander. While the oysters are draining, prepare the brine.

2 quarts water	1 cup brown sugar
1 cup soy sauce	1 tsp. onion powder
1/2 cup salt	1 tsp. garlic powder

Stir the ingredients into the water, simmer, and stir until all ingredients are dissolved. Bring to a boil and add the oysters. Cook for 10 to 30 minutes. Remove the cooked oysters and drain in a colander.

Hot-smoke the oysters in a smoker for 4 to 6 hours for added flavor.

Smoked Smelt

Smelt are excellent smoked and make great appetizers. Cure and smoke same as for larger fish, but do not brine very long, about an hour or two, and then wash thoroughly in fresh water. Smoke the smelt until they turn a golden brown and are completely cooked. Allow to cool before serving. The bones and all are eaten and make a delicious appetizer or meal.

Smoked Eel

A delicious but, granted, fairly localized, smoked product, eel can be skinned and filleted for smoking, but it is traditionally smoked whole. Freshly killed eels should be washed in clean, cold water and scraped to remove slime. You may need to repeat this step several times to remove all traces of slime. Another method to deslime eel is to simply cover them with salt in a noncorrosive container. Eviscerate the eel and remove all traces of blood. Soak the eel in 80 percent brine for 30 minutes to 1 hour. Allow to dry thoroughly and then hang from rods in the smoker. Place small sticks between the belly flaps to keep them open. Smoke at 100°F to 120°F for 1 hour, raise the temperature gradually to 140°F, and then gradually to 180°F or 200°F, and smoke until an internal temperature of 160°F is reached.

Pickled Fish

Although salt is applied, pickling fish is not the same as curing in brine and then drying. Almost any type of fish can be pickled, but traditional favorites include herring and carp. Pickled fish is not cooked and should be frozen at 0°F for two weeks or longer to get rid of any parasites. Check your freezer to make sure the temperature is cold enough. Pickled fish is also only mildly preserved and must be kept refrigerated.

Eviscerate and wash the fish in clean, cold water. Remove fins, tail, and head. Scale or skin, then cut into 1-inch-thick pieces. Soak the pieces in brine made of 1 gallon of water with 2 cups of salt dissolved in the water.

While the fish are soaking, make up a pickling mix (leaving out the onion). Use the proportions below to make up a pickling mix sufficient to cover the fish. Bring the mix to a boil and set aside to cool before using.

1 1/2 quarts distilled white vinegar	6 bay leaves
2 1/2 pints water	1/4 cup reconstituted lemon juice
1 tsp. whole allspice	1/2 cup brown sugar
1 tbsp. mixed pickling spices	1 large onion (sliced)
1 tsp. mustard seed	

Drain fish from brine and pack in clean, sterilized glass jars. Place onion slices in jars and pour cooled pickling mix over. Place lids on the jars and shake well to distribute the spices. Keep refrigerated and use within 6 to 8 weeks.

⑨
Wild Game Recipes

FOR MANY HUNTERS, salt curing and smoking is a favorite means of preserving or, more commonly, cooking game, including venison, big game, and game birds. As with other types of meats, different methods of cure application can be used, depending on the cut of meat and the desired end result. This includes dry cure and injection cure or stitch or artery pumping, or a combination of the methods. Small cuts, such as loins and cuts from the hams and shoulders, are commonly cured using the dry-cure method. Larger pieces, such as hams, are best cured with a combination of dry cure and injection. Salt-cured game meats may or may not be smoked, but all must be hot-smoked or cooked. As with all meats, it's important to start with clean, fresh, safe, and well-chilled meat.

Corned Deer Loin

The loin is one of our favorite cuts to corn. It tastes almost like corned beef, with a mild game flavor, and there is no fat to be removed. It is, however, a bit dryer because there are no connective fat tissues or marbling. This recipe is easy using Morton Salt Tender Quick.

5 lbs. deer loin, boneless venison roast, or combination	6 tbsps. Morton Tender Quick
	1 tbsp. garlic powder
2 to 3 quarts water, to cover	1 tbsp. onion powder
1 cup brown sugar	2 tbsps. mixed pickling spices

Place the brine ingredients in a large saucepan. Bring to a boil, reduce to simmer, and stir until all ingredients are dissolved. Place in a large food-grade plastic container and place in the refrigerator to chill. Once chilled, add the chilled meat, cover the container, and keep refrigerated 5 to 7 days, depending on the thickness of the meat. Turn the meat once each day. Remove, wash in cold running water, and place in a large pot. Cover with water and simmer for 3 to 4 hours. Remove, chill, thin-slice, and serve. This makes an excellent Reuben sandwich. Freeze excess for future use in vacuum-seal bags.

Venison Ham

A cured ham from a deer is mighty tasty. The cured deer ham can be smoke-cooked, oven-roasted, or boiled. A deer ham is best cured using the combination dry cure and injection. The same basic recipes and techniques for curing a pork ham can also be used on a deer ham. In this case, the deer ham is cut from the carcass by making a cut similar to that for long-cut pork ham, cutting through the aitchbone and pelvic arch.

Trim and smooth up the butt end and cut off the shank fairly short, as the shank of a deer has little meat. A simple method is to use Morton Salt Sugar Cure (Plain) mix. Make up the sweet pickle cure first by combining one cup of the Morton Salt Sugar Cure (Plain) mix with 4 cups of clean, cool water and mix until dissolved. Weigh the ham and make up enough pickle to have 1 ounce of pickle for each pound of meat. Pump the pickle into the ham, using a meat pump and injecting along the bone structure. After the ham has been pumped, apply a dry cure to the meat surface. Again, Morton Salt Sugar Cure (Plain) mix makes the chore easy. Since deer hams are usually not aged, use 1 tablespoon of mix per pound of meat. Measure out enough cure mix to do the ham and divide into 3 equal parts. Apply 1/3 of the mix on the first day and place in a large covered food-grade plastic container or

BELOW: A ham from a deer can also be salt-cured just like a pork ham. The best method is the combination- cure method of injecting a pickle cure along with using the dry- cure method.

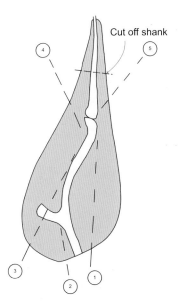

Cut off shank

bag and then place in a refrigerator with a temperature of 38°F. After 7 days, apply another third of the cure, and then on day 14, apply the last third. The deer ham should be cured after 14 to 15 days, followed by an equalization period.

To equalize, remove from the container and soak the ham for about an hour in clean, cold water. Lightly scrub off the excess cure and place in a clean food-grade plastic bag or container. Place in the refrigerator and allow the salt to equalize throughout the ham. This will take about 14 days. Remove from the bag, soak again in cold water, and then hang to dry. The ham can be smoked or cooked, but the ham must be heat-treated to 160°F internally before being consumed.

Bacon Smoked Quail

If you're looking for a unique quail dish, this is it. The same recipe can be used with pheasant, quail, or chukars. First step is to make up a weak brine solution.

1 quart water	1 tsp. onion powder
3 tbsps. noniodized salt	1 tsp. garlic powder
1 tbsp. lemon juice concentrate	1 tsp. black pepper
1 tsp. dried parsley	1/2 tsp. paprika

Dissolve the salt and spices in the heated water, then place the brine in the refrigerator to cool. Place the cleaned whole birds in a food-grade plastic bag and pour the mixed (cold) brine into the bag. Place in a plastic food container and place in a refrigerator. Allow to cure for 4 to 6 hours, turning frequently to make sure all pieces are thoroughly soaked.

Remove, rinse the brine, and place on paper towels. Pat the birds dry with paper towels. Drape bacon pieces over the breast of the birds and pin in place with toothpicks. Smoke using apple or alder at 200°F to 225°F until the internal temperature reaches 165°F.

ABOVE: Upland game—such as quail, pheasant, and chukars—can all be lightly brined and smoked. Clean thoroughly, removing all feathers and pinfeathers. Cut off feet. Dig out any shot and cut away any bloody areas.

BELOW: Cure upland birds in a mild brine for 4 to 6 hours, then hot-smoke in a smoker with a water pan to an internal temperature of 170°F. Bacon strips added across the breast can add moistness.

Apple-Soaked Mallard Breasts

Just about any duck can be used—including teal, wood ducks, and gadwalls, one of my favorite eating ducks. The bigger ducks provide a more moist breast, and of course, more meat. Breast out the ducks and remove the breast skin. Cut away any bloody areas and dig out any shot. Steel shot can be mighty hard on the teeth. Prepare the following brine, and soak the breast overnight. This will brine about 2 pounds of duck breasts, or the breasts from 2 mallards.

1 quart water	2 bay leaves (crushed)
1 can or jar applesauce	1 tsp. garlic powder
3 tbsps. noniodized salt	1 tsp. onion powder
1/4 to 1/2 cup brown sugar	1 tbsp. lemon pepper

Thoroughly mix all ingredients, except for the lemon pepper, in hot water and allow the brine to cool. Place duck breasts and brine in a food-grade plastic container and leave overnight in the refrigerator. Remove from the brine, rinse, and dry with paper towels. Sprinkle each duck breast with lemon pepper and place bacon strips across the breasts. Secure in place with toothpicks. Smoke with applewood in a smoker at 225°F for a couple of hours. Duck is normally best eaten medium rare, but can be smoked until the internal temperature reaches 165°F. To keep duck breasts moist, place in aluminum foil after the first hour of smoking. Slice and serve with baked apples.

Smoked Sky Carp

Snow geese are called sky carp for good reason—they're plentiful, with large limits, tough to bag, and tough to cook. The meat is extremely dark and dry. Salt curing and smoking can, however, provide delicious goose meat, and is a good choice for the big Canada honkers as well. Only the breasts are used. Remove the breasts and remove the skins.

Clean up any bloody spots and remove all steel shot. Slice the breasts into 1/2-inch strips. Make the following brine and soak the strips overnight in a refrigerator.

1 cup soy sauce	2 tbsps. Worcestershire sauce
1 cup water	1 tsp. garlic powder
1 cup brown sugar	1 tsp. onion powder
2 tbsps. noniodized salt	1 tsp. ground red pepper, or to taste
1 tbsp. black pepper	

Remove the strips, rinse under cold water, and dry on paper towels. Smoke using hickory at 200 to 225 degrees for a couple of hours, or until the internal temperature reaches 170°F. After the first hour of smoking, place the goose strips in aluminum foil to prevent drying.

Wild Game Summer Sausage

Summer sausage is an excellent way of salt-curing and smoking wild game meats. Although venison is probably the single most common meat used, just about any big game, small game, and waterfowl can be used to make summer sausage. The recipe used is easy, with Morton Salt Tender Quick mix. You will need some pork trimmings to add a little fat to the wild game meat. This is a good recipe to use with electric smokers.

3 lbs. venison or other wild game	1/2 tsp. ground ginger
2 lbs. pork trimmings	1/2 tsp. ground mustard
1 tbsp. black pepper	1 tsp. garlic powder
5 tbsps. Morton Tender Quick mix	4 tbsps. corn syrup
1 tsp. ground coriander	

Weigh the meats separately. Cut chilled meat into 1-inch cubes and grind through a 3/16-inch grinder plate. Mix the dry spice ingredients in a glass bowl and sprinkle over the ground meat. Dribble the

corn syrup over the ground meat and thoroughly mix all. Place in a plastic or glass bowl and refrigerate overnight. Spread the meat out to about a 1-inch depth in a shallow, flat pan and freeze for an hour or so or until the meat is partially frozen. Remove and regrind the partially frozen meat through a 1/8-inch plate. Stuff the ground meat into synthetic casings. If your meat grinder has a stuffing attachment, the final grinding and stuffing can be done in one step.

Hang the stuffed casings on drying racks and dry at room temperature for 4 to 5 hours, or hang in a smoker on sticks, with the damper open until the casings are dry to the touch. Set the temperature of the smoker to 120°F or 130°F, add wood chips, and smoke for 3 to 4 hours. Raise the temperature to 170°F and cook until the internal temperature reaches 165°F. Remove from the smoker and shower the casings with cold water. Place back in the cooled-down smoker and hang at room temperature for 1 to 3 hours or until dry. Freeze sausages that will not be consumed within a week. For more sausage recipes, see Monte Burch's *The Complete Guide to Sausage Making*.

PART 2

Jerky

① Old-Time Jerky Making

NO MATTER WHAT the main ingredient was or is—mastodon, elk, deer, African or Australian game, beef, fish, you name it—the old-fashioned method of making jerky has been around for a long time. Old-time jerky is still easy to make and still provides a great food source. In the old days, jerky making was very simple. The Native Americans simply cut thin strips of meat from game they had killed, then hung the strips over racks made of thin branches. In the dry Southwest and the Plains, meat dried quickly and easily with the use of this method. In the North, a small, smoky fire was often used to speed the drying process. Not only did this help the drying process, but it also kept away the blowflies. In the Northwest, smokehouses were constructed to protect the meat and aid in the drying process. If the Native Americans had access to salt, it was applied as well. The Native Americans also dried salmon, placing them on long racks as they removed fish from the fish wheels in the rivers.

One of my favorite outdoor writers from earlier days was Colonel Townsend Whelen. This is his description of jerky making:

The old-time method of jerky making using only the sun has been a tradition in many cultures, including those of the American West.

Jerky is lean meat cut in strips and dried over a fire or in the sun. Cut the lean, fresh red meat in long, wide strips about half an inch thick. Hang these on a framework about 4 to 6 feet off the ground. Under the rack, build a small, slow, smoky fire of any non-resinous wood. Let the meat dry in the sun and wind. Cover it at night or in rain. It should dry in several days.

The fire should not be hot enough to cook the meat since its chief use is to keep flies away. When jerked, the meat will be hard, more or less black outside, and will keep almost indefinitely away from damp and flies.

It is best eaten just as it is; just bite off a chunk and chew. Eaten thus, it is quite tasty. It may also be cooked in stews and is very concentrated and nourishing. A little goes a long way as an emergency ration, but alone it is not good food for long, continued consumption, as it lacks the necessary fat.

Following is a campsite jerky technique I learned from an old-time Wyoming big-game guide. He described his method of making jerky to me as we chewed on some while glassing for elk:

Cut the meat into strips, lay on a flat surface, and sprinkle both sides with black pepper. Lightly sprinkle with salt. Rub the salt and pepper well into all sides of the strips. Cut holes in the ends of the strips and thread white cotton or butchers cord through each hole, tying off into loops. Bring a pot of water to boil and immerse the strips into the boiling water for 15 to 20 seconds, remove, then re-dip.

The traditional Native American method of drying meat for jerky consisted of hanging meat strips over racks made of thin branches. A small smoky fire under the meat not only kept away insects but also added flavor and aided in the drying process.

One old-time method of preheating jerky strips was to place loops of string through holes cut in the ends of the strips. These were threaded onto a stick and dipped in a pot of boiling water.

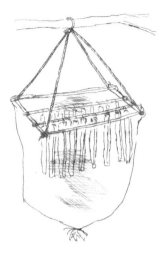

A bag or "tent" of cheesecloth was often used to help keep off insects while the jerky dried.

Hang the strips to dry. If the strips are hung outside in the sunshine, cover them with a cheesecloth tent to keep off insects and make sure they're high enough so dogs and other critters can't get to them. The strips can also be hung on clothesline in a cold, dry room. The strips should be dry in 4 to 5 days.

Another traditional method involves the use of curing salt, an old-time product. It's easy to make your own curing salt. Take 1 pound of canning salt, 6 ounces of Prague powder, 3 ounces of sugar, and 2 ounces of white pepper. You can substitute brown sugar and black pepper. If you like hot jerky, add ground red pepper or cayenne pepper to suit. Mix all together and rub the mix over all the meat slices. Leave in a cool place overnight, then dry. In damp weather, the slices can be dried in an oven or meat smoker.

An old-time oven method is to lay strips in a glass dish, place a drop of Liquid Smoke over each strip, and use a pastry brush to evenly coat each strip. Sprinkle seasoning salt and seasoned pepper over the layer. Add a light sprinkling of sugar and garlic powder if you like garlic. Add another layer of strips, brush with Liquid Smoke, sprinkle with salt and pepper, then add another layer of strips. Continue adding and seasoning until the dish is full or you run out of strips. Cover the dish and set in a refrigerator or cool area (below 40°F) overnight. Dry in an oven set to 200°F or in a dehydrator.

Pemmican

Made from jerky, pemmican was also a staple food of the Native Americans. Another of my favorite old-time writers, George Leonard Herter, in his *Professional Guides' Manual*, published in 1966, stated, "Pemmican properly made is one of the finest foods that you can take into the wilderness or for a survival food in case of atomic bombing. Pemmican keeps indefinitely. Today, in our wonderful atomic age, pemmican is part of the survival ration of the newest United States Air Corps jet bombers."

According to Col. Townsend Whelen, "To make pemmican you start with jerky and shred it by pounding. Then, take a lot of raw animal fat, cut it into small pieces about the size of walnuts, and fry these out in a pan over a slow fire, not letting the grease boil up. When the grease is all out of the lumps, discard these and pour the hot fat over the shredded jerky, mixing the two together until you have about the consistency of ordinary sausage. Then, pack the pemmican in waterproof bags. The Indians used skin bags."

The proportions should be about half lean meat and half rendered fat. The Native Americans also added fruits such as wild grapes, dried berries and beans, corn, herbs, and other items. These added vitamin C, which prevents scurvy, as well as other nutrients and gave the pemmican different tastes. To use, place the dried block of pemmican in water and bring to a boil. Herter suggests dropping in some chili powder, soaking some beans overnight, adding them, and then "You will have an excellent chili con carne."

If you want to try making pemmican, the following is a recipe to make approximately 10 pounds.

5 lbs. jerky	3/4 lb. raisins or dried currants
1/2 lb. brown sugar	4 lbs. melted fat

Pound the jerky until it crumbles, and mix all ingredients together.

If you want to make a more modern version, first run the jerky through a food processor. Then, add a cup of raisins, a cup of salted peanuts, and a cup of brown sugar for each pound of jerky.

Jerky was often made into pemmican by the Native Americans, utilizing suet cooked to render off the fat, then adding the fat, as well as fruits, herbs, and other ingredients to jerky that had been pulverized. Today you can use a food processor to quickly pulverize the jerky.

Other dried fruits such as cranberries can also be used. Sugar is optional, a matter of taste. The sugar can also be replaced with chocolate or any other flavor of chips (butterscotch, semisweet, milk chocolate, and so on). Press the mixture into a pan, packing tightly. Pour melted suet or other fat over the mixture, using only enough fat to hold the ingredients together. It's easy to get too much fat. A modern alternative to melted suet or bacon grease is a butter-flavored shortening. Allow the mixture to cool and then cut into squares for storage and use.

To make a chili version, leave out the sugar and dried fruit and stir in chili seasoning with the ground jerky and fat. To use, add a chunk of the chili-flavored pemmican to a pot of cooking beans. This makes a very hardy camp meal.

Add raisins or other dried fruits and nuts to the pulverized jerky, then stir in the warm, melted fat, adding only enough fat to hold the mixture together. A butter-flavored shortening is a good alternative to bacon grease or rendered suet.

For long-term storage, it's best to keep your jerky supply in the freezer and make pemmican just before consuming. Except for short periods of time, keep pemmican in the refrigerator, especially in warm weather.

It's possible to make jerky and pemmican using these age-old traditional methods, even in a remote camp, but always follow the safe food processing issues discussed in Chapter 3.

② Sliced Muscle-Meat Jerky

THE SIMPLEST, TRADITIONAL method of producing jerky is to use sliced muscle meat. Almost any type of big-game meat can be sliced into jerky, including wild game such as deer, elk, moose, sheep, and antelope. Other meats, such as wild turkey breast, can also

Traditional jerky was, and still is, made by cutting muscle meat into slices. Wild game such as deer and elk were and are popular meat choices. Beef is a traditional meat, while the Native Americans utilized buffalo.

be made into jerky slices. Beef is a traditional sliced-jerky meat in many parts of the world, and buffalo was a favorite of early settlers and Native Americans.

In the 1970s I was a young freelance writer and book author with a family to feed. Since we lived on a farm in the Ozarks, venison was a regular food staple. I started making jerky in the old-fashioned, sliced muscle-meat method back then and have been doing so ever since. With several deer usually taken each season, most of one deer would end up being turned into jerky.

Trimming and Slicing

Traditionally I've used the hams or rear quarters of venison for old-fashioned, sliced jerky. The hams do provide good meat cuts for other uses, including roasts and hamburgers, but they also slice extremely well for jerky. Then, I use the butcher trimmings, left from cutting the sliced jerky, for ground meat, both for jerky and to cook in chili, tacos, meatloaf, and so forth. Once the rear quarters have been deboned from the carcass, cut the meat into long chunks by simply following the muscles where they are separated by sinew. Cut away any fat, as fat not only doesn't dry properly, but it adds a gamey flavor to the meat. Also cut away as much sinew as possible. Next, using a sharp butcher knife, slice the meat into thin strips, following the grain, not across it. Slice into strips either 1/8- or 1/4-inch thick. Strips can actually be cut as thick as 3/8 inch, but any thicker and they become extremely hard to dry. The thickness is actually only a matter of taste. It takes longer to dry and cure the thicker strips, but they also tend to turn out less brittle when dried.

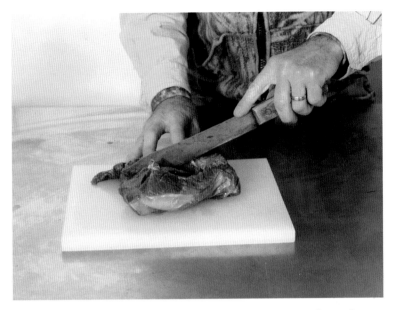

Once the meat has been deboned from the carcass, cut it into long chunks, following the sinew or natural muscle divisions.

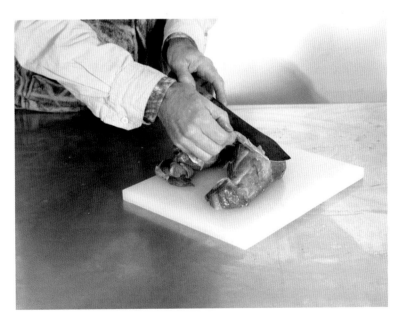

Cut away all fat and sinew.

Then, slice into strips of the desired thickness.

The Hi Mountain Jerky Board and stainless steel knife makes it easier to more consistently slice meat to the correct thickness and therefore results in a more consistent drying of the jerky. If a strip of meat is 1/8 inch at one end and 3/8 inch at the opposite end, for instance, it will be difficult for it to dry evenly. The board comes with a lip around the edge that allows you to slice 1/8- or 1/4-inch thicknesses. Place the meat on the board and slice holding the knife with the flat side of the blade down on the edge lips.

The simplest and most consistent method of slicing meat is with an electric slicer. Regardless of the method used, for easier slicing, lay the trimmed meat pieces on a cookie sheet or other flat surface, place in the freezer until the meat is partially frozen (usually about an hour to hour and a half) and then slice.

The Hi Mountain slicing kit includes a wooden board with raised edges to help guide the stainless steel knife (included) for precise 1 8- or 1 4-inch slices.

Place the meat to be sliced in the board, and slide the knife (flat side down) along the raised edges.

Frequently dipping the knife blade in cold water assists in slicing.

An electric slicer can be used to cut precise thickness slices fast.

Curing

Jerky can be nothing more than dried meat using a variety of heat and drying sources. Curing the jerky meat, however, not only adds to the preservation but improves the flavor. Jerky can also be made with a wide range of flavorings, depending on the seasonings and spices used. You can make up your own curing recipes, or use some of the wide range of cures and seasoning mixes available commercially. The latter often contain nitrites to aid in curing. Make sure you follow the recipe and use the proper amount of cure so you don't have too little or too much nitrite. In order to use any recipe, homemade or commercial, you need to know the exact weight of the sliced meat. Regardless, the commercially prepared cures and mixes allow you to make your own jerky with proven taste results. On the other hand, experimenting with homemade recipes is half the fun of making jerky.

The meat can simply be dried, but in most instances, a cure or seasoning is added for flavor and can also be used to help cure the meat. You can make your own cure and seasonings by using a variety of curing agents and spices.

For most cure and seasoning mixes, you must know the exact amount of meat you're dealing with. We use our postal scale to weigh up to 10 pounds of meat.

Curing can be accomplished by using one of two methods: marinating the meat in a liquid marinade or using a dry rub to coat the meat. In most instances, the meat must sit a length of time before drying in order for the cure to work through. I've used both methods with good results.

The following recipes are primarily for venison, but elk, moose, antelope, or beef could be substituted.

Marinades

Lawry Marinade

2 lbs. meat, thinly sliced	2/3 cup Lawry's Seasoned Marinade or Mesquite Marinade

This is one of the simplest recipes I've used. Place 2/3 cup of Lawry's Seasoned Marinade or Mesquite Marinade and 2 pounds of meat strips in a resealable plastic bag or sealable plastic container. Mix until all meat strips are evenly coated. Place in the refrigerator for 2 to 4 hours. Remove, allow the meat strips to drain on paper towels, and then pat the surface dry with more paper towels. Sprinkle on Lawry's Seasoned Pepper or Garlic Pepper.

To dry, turn on an oven to 200°F, prop the door open an inch or so, and dry the meat. Drying can take anywhere from 2 to 6 hours. Check often for doneness. Jerky strips should feel like leather and bend but not break. The meat should be dried but not cooked. Store in an airtight container and keep in a cool, dry place.

Burch Marinade No. 1

2 lbs. meat, thinly sliced

1 cup of soy sauce

1 tbsp. garlic powder

1 tbsp. onion powder

Water to cover

The meat can be cured in one of two methods. Using a liquid marinade is very common.

The dry cure and spices are mixed with enough water or liquid to cover the meat.

Allow the meat to soak in the liquid in a nonmetallic container, usually overnight.

I've used this simple marinade for almost 40 years. Many recipes contain salt as a curing agent, but other ingredients can be used to replace the salt. Soy sauce is substituted for the salt in this recipe. Mix the ingredients; place them with the meat in a glass bowl, plastic container, or resealable plastic bag; and refrigerate for 12 hours or overnight. Remove, drain, and pat dry. Dry in the oven or a dehydrator. If you desire a spicier taste, add Tabasco Sauce.

Burch Marinade No. 2

2 lbs. meat, thinly sliced	1/2 tsp. black pepper
Water to cover	2 tbsps. onion powder or one large, fresh
1 tbsp. Worcestershire sauce	onion, finely diced
1 tbsp. salt	1 tsp. garlic powder
1/2 cup brown sugar	Tabasco sauce to taste

Jerky is marinated to provide the taste and can be made as spicy or as mild as you prefer. Above is a very mild recipe we've used for years. Once the meat is sliced and all fat removed, place it in freezer bags, freeze for 60 days at 0°F to 5°F, and then cure and dry in an oven or dehydrator.

Place the thinly sliced meat in a bowl, cover with the marinade, and refrigerate for at least 24 hours. Stir occasionally. You can make the marinade spicier by adding soy sauce or teriyaki sauce and hotter by adding ground red pepper. Some folks like to add more sugar. You can also simply sprinkle additional spices onto the meat before drying, if you want even more flavor.

Dry Rub Cures

Dry rub cures normally consist of the dry ingredients sprinkled over and rubbed into the meat. The rub ingredients are then allowed time to work into the meat before drying.

Easy-Does-it Oven Jerky

Meat, thinly sliced   Seasoned salt
Liquid Smoke

This is a great recipe if you just want to try your hand at making jerky without a lot of hassle and ingredients. It's quick, easy, and tasty. All you'll need is meat, Liquid Smoke, and seasoned salt. Slice the meat into 1/8-inch-thick pieces. Brush a bit of Liquid Smoke on both sides of each piece and then dust each piece with seasoned salt. Place the coated meat strips in a covered container or sealable plastic bag and store in a refrigerator. Allow the strips to marinate overnight. Remove and pat dry with paper towels to get rid of any excess moisture. Place cookie sheets or aluminum foil on the bottom of the oven to catch any dripping. Spray the oven racks with cooking oil and then hang the meat slices over the racks. Set the oven to 200°F and roast the meat until liquid begins to drip. Reduce the heat to 140°F, and dry the meat with the oven door slightly open. The jerky should be ready in 4 to 6 hours, but test frequently.

Morton Salt Jerky

2 lbs. lean beef or game, sliced   2 teaspoons sugar
2 tablespoons Morton Tender Quick mix   1 teaspoon ground black pepper
    or Morton Sugar Cure (Plain) mix   1 teaspoon garlic powder

One simple and easy dry-rub cure recipe I've used with extremely good results is the Morton salt jerky recipe. It is extremely mild yet long lasting.

In a small bowl, stir together the Morton Tender Quick mix or Morton Sugar Cure mix and the remaining ingredients. Place the meat on a clean surface or on a large flat pan, and rub all surfaces with the cure mixture. You can also put the ingredients in a bowl, and then rub the cure into each slice of meat. Place the rubbed strips in a plastic food storage bag, and seal shut. Allow to cure in the refrigerator for at least 1 hour. After curing, rinse strips under cold running water and pat dry with paper towels. Arrange meat strips on a single layer on greased racks in a shallow baking pan. Meat edges should not overlap. Place in a 325°F oven, and cook meat to an internal temperature of 160°F. Dry meat in a home dehydrator following the manufacturer's instructions.

A dry-rub technique can also be used. Mix the ingredients thoroughly.

Sprinkle the ingredients over the meat strips and toss to coat well, or rub the mix into the meat surfaces.

Place the rubbed strips in a glass or plastic container or zippered plastic bag to cure.

Venison Jerky—Chef Williams Style

2 lbs. venison roast	Vegetable oil
Cajun Injector Wild Game Marinade	Cajun Injector Cajun Shake
Cane Syrup Recipe	

One of the most unusual jerky recipes is a combination of marinade and dry rub. This recipe is from Chef Williams, the founder of Cajun Injector Marinades. We found the recipe in an old pamphlet from Cajun Injector and have used it for years. The company doesn't make the Cane Syrup Recipe Marinade anymore, but their other marinades, such as Teriyaki and Honey, also work well with this recipe.

Inject the roast using 2 ounces of Cajun Injector Wild Game Marinade per pound of meat. Slice the roast into 1/8- to 1/4-inch slices, slicing with the grain. Sprinkle with Cajun Shake. Brown the venison strips in hot oil in a large skillet, turning as the strips brown. Place in a dish and cover the strips with s ounces of Cajun Injector Wild Game Marinade and sprinkle with the Cajun Shake. Let the strips sit overnight in a refrigerator. Drain off the marinade and pat dry. Dry in a dehydrator or bake in a 150°F oven with door partially open for s hours or until the meat is dry.

Using Prepared Cures and Mixes

I've experimented with a number of prepared cures and mixes and found them easy and fun to use. In all cases, the product comes in two packages: a cure and a seasoning. It's extremely important to follow the mixture amounts suggested by the manufacturer for the specific poundage of meat. It's also a good idea to make up small batches of jerky to test for taste. The following are jerky-making suggestions and methods from the products I've tested.

Uncle Buck's Jerky Regular Seasoning and Cure

Up to 10 lbs. meat	8 teaspoons seasoning for every 2 lbs. meat
2 cups water for every 2 lbs. meat	½ teaspoon cure for every 2 lbs. meat

Cut the meat strips 1/8-inch-thick and 8 inches long. To cure up to 10 pounds of meat, mix the cure packet and seasoning with 5 cups of water. For smaller batches, mix 8 teaspoons seasoning, ½ teaspoon of cure, 2 cups of water, and 2 pounds of meat strips. Mix well and marinate strips in solution for 8 hours under refrigeration. Hang strips in oven at lowest setting with door open to the first stop until jerky reaches desired dryness. A dehydrator may also be used. This product is also available in hickory and mesquite flavors.

Uncle Buck's Jerky Seasoning and Cure provides a great way of making jerky with a sure-fire recipe.

Place the cure mix in water.

Add the seasoning mix and stir well.

Add the strips.

Stir well and refrigerate for 8 hours.

Eastman Outdoors Jerky Cure and Seasoning

Lean meat

3 teaspoons seasoning for every 1 lb. of meat

1 teaspoon for every 1 lb. of meat

Cure

Start with the leanest meat possible. Trim excess fat and partially freeze the meat for easier slicing. Cut along the grain into strips no more than 3/8-inch thick. Weigh the strips to determine how much seasoning and cure to use. Then follow the mixing chart using standard measuring spoons and a nonmetallic bowl, mixing 3 teaspoons of seasoning and 1 level teaspoon of cure for each pound of meat. Gently toss the strips with the mixture. For best results, use the Eastman Outdoors Reveo to infuse maximum flavor into the meat. Cover the bowl, or place strips in a sealable plastic bag, and refrigerate for at least 24 hours. Use an oven, smoker, or dehydrator to dry. The Eastman Outdoors products are available in original, hickory, teriyaki, mesquite, and whiskey pepper.

Hi Mountain Jerky Cure and Seasonings

Hi Mountain Jerky Cure and Seasonings are old recipes, made without preservatives, and the dried jerky must be kept frozen or refrigerated. The company suggests you have fun with jerky. If you have meat that has been in the freezer for a while, for example, don't waste it: make it into jerky instead. Or, if you have a roast, try cutting it into 1-inch squares, and make jerky nuggets. Hi Mountain products are available in ten authentic Western recipes: original blend, pepper, mesquite, hickory, mandarin teriyaki, bourbon BBQ, cajun, cracked pepper and garlic, inferno, and pepperoni.

Cut the meat into strips of desired lengths and widths, allowing for shrinkage. Weigh the meat after cutting into strips and trimming. Now you know the exact amount of mix to use. Mix the spices, and cure according to the spice and cure mixing chart. Mix only the amount you need. Be sure to store the remaining unmixed spices and cure in an airtight container until needed. Hi Mountain recommends that you always make sure you mix the cure and seasonings exactly and correctly. Fluff the cure and seasoning before measuring. Always use standard measuring spoons, level full. Scrape off excess cure or seasoning with a table knife, leaving the measuring spoon level. Do not compact.

Hi Mountain provides a wide variety of flavors in their line of cure and seasoning mixes that can be used with sliced jerky.

The cure and seasoning spices must be mixed together precisely according to the amount of meat being treated.

Hi Mountain provides a shaker bottle to apply their mixes. Apply to one side of the meat, turn, and apply to the opposite.

Thoroughly mix the meat and spices in a nonmetallic bowl.

Place the slices in a nonmetallic container or zippered plastic bag and refrigerate for the time specified by the recipe.

Lay the strips flat on an even surface. (If you have just washed the game meat, be sure to pat it dry before applying the cure and seasoning.) Using the blended spices and cure, apply to the prepared meat using the shaker bottle enclosed with the mixes. Sprinkle the first side of the meat strips with approximately half of the measured mix. Turn the meat strips over and sprinkle the remaining mix on the meat strips. If you can't get an even distribution on the meat, especially on the ends and edges, put all the seasoned strips in a large mixing bowl and tumble by hand until the cure and seasoning have been spread evenly over the entire batch.

Stack the strips, pressed together tightly, in a nonmetallic container or a resealable plastic bag and refrigerate for at least 24 hours. Hi Mountain Jerky Cure and Seasoning is specially formulated to penetrate meat at the rate of 1/4 inch per 24 hours. If thicker pieces of meat are used, increase curing time accordingly; for instance, cure 3/8-inch-thick strips for approximately 28 hours. Do not cure any meat less than 24 hours.

Drying

The following instruction for cooking or smoking jerky is from Hi Mountain: "Place foil or pan on bottom of oven to catch drippings. Lay the strips on the oven racks, making sure there is air between each piece (our Jerky Screens are perfect here). Place in the oven for 1 to 1¼ hours at 200°F with the oven door open just a crack. Taste the jerky frequently. When the jerky is cooked to your liking, stop cooking. The jerky is made with cure and seasonings, it does not have to be dry to the point where you can't chew it like store bought jerky. Remember, taste often while cooking or smoking."

Our oven door won't stay open just a crack, so I keep the door open with a wooden clothespin. A wooden stick of the desired thickness would also work. Another easy oven-dry technique is to insert toothpicks in the ends of the meat strips and hang the strips from the oven racks. This works, but they also fall easily, especially when you're trying to move the racks in and out of the oven.

If using a smokehouse or smoker, Hi Mountain recommends experimenting. All home smokers are different in size, wall thickness, location (inside or outside), outside temperature, wind, heat source, and so forth. Hi Mountain recommends smoking the jerky at 200°F for 1½ to 2 hours with smoke on; however, if your smoker will not reach 200°F, leave the meat in longer. Do not smoke for more than

The jerky strips can be dried in several ways. The first step, however, is to place the marinated or dry-rub strips on paper toweling and pat dry.

One of the simplest drying methods for the home jerky maker to use is drying in a kitchen oven with the door propped open slightly. Shown here are jerky strips on Hi Mountain jerky racks.

Jerky strips can also be suspended on ordinary oven racks using toothpicks threaded through holes in the ends of the strips.

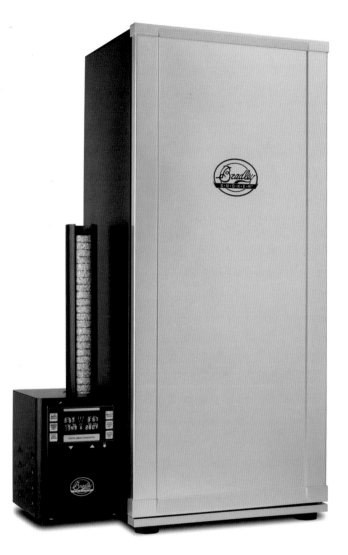

Electric smokers that will reach temperatures of at least 140°F are also good choices for drying jerky strips.

3 hours until you have tasted the first batch. Do not overcook, and do not oversmoke. Too much smoke can produce a bitter flavor.

If using a dehydrator, Hi Mountain recommends that you follow your dehydrator instructions. Again, jerky does not have to be cooked so hard that you can't chew it. Test it frequently.

A food dehydrator that will reach a temperature of 140°F is an excellent tool for dehydrating and drying jerky.

③

Small Game and Wildfowl Jerky

UPLAND GAMEBIRDS, WATERFOWL, and small game can all be made into jerky. The type of jerky—muscle meat or ground meat—depends on the species being used. Small game such as rabbits, squirrels, groundhog, and others, as well as ducks and geese, are best made into ground-meat recipes. Small game often don't provide large enough cuts of meat for slices. The dark meat of ducks and geese, especially snow geese, usually tastes better when ground with other, milder meats or when made into a spicy ground-meat recipe. In fact, a great way of cleaning out your freezer is to make up a smorgasbord

Geese, ducks, and small game can all be made into excellent jerky.

of ground-meat jerky from whatever overabundance of game meat you have on hand at the end of the season.

Ground-Meat Smorgasbord Jerky

2 lbs. wild game meat, ground	1 tablespoon soy sauce
2 tablespoons Morton Tender Quick	1/2 cup water
1 teaspoon lemon pepper	1 to 4 drops hot sauce (or more if you
1 teaspoon onion powder	prefer more heat)
1 teaspoon garlic powder	

As with all other types of meat, make sure you use safe field dressing and butchering processes, especially when making ground-meat

The first step is to debone the birds and small game. Then, grind the meat. Add the cure and seasoning, refrigerate, and dry.

jerky. Pathogens can easily be spread throughout the ground meat. Also, make sure you use meat only from healthy animals. Debone all the meat from the game animals or birds, and remove all fat and gristle. Cut away all bloody meat, and remove any shot from gunshot game. Ground-meat smorgasbord jerky is a case where you can use even the tougher, lesser cuts, including the thighs of birds such as wild turkeys or geese. If you use the legs, make sure you remove all the tiny, flexible tendons and bones. It's a good idea to first soak all the deboned pieces in salt water overnight in a refrigerator. This will help remove some of the gaminess and also tenderize some of the tougher pieces. If the meat hasn't been frozen, freeze it for 60 days. Partially thaw the meat, then grind it.

Mix the ingredients together, making sure the spices are dissolved in the liquid. Pour the seasoning and cure over the ground meat in a nonmetallic bowl, and mix well until tacky. Refrigerate overnight, extrude or form, and then dry. This seasoned and cured ground meat can be made into an excellent snack stick.

Turkey Jerky

Domestic and wild turkeys are large enough to be made into jerky in the sliced-muscle-meat method. Only the breast meat is used for this procedure. As with other types of muscle-meat jerky, remove all fat and connective tissue. Also, remove the skin. Then, slice into 1/4-inch-thick strips. Partially freezing to firm the meat will help with slicing it into uniform, thin strips. Some of the best jerky I've tasted is the Hi Mountain Turkey Hunter's Bourbon Blend Jerky Cure and Seasoning. After cutting the meat into strips, weigh the meat so you know the exact amount of cure and seasoning mix needed. Mix the spices, and cure according to the mixing chart (per weight) included with the instructions. Mix only the amount needed. Be sure to store the remaining unmixed spices, and cure in an airtight container until needed. Lay the strips flat on an even surface, and pat dry with a paper towel. Apply the mixed spices and cure to the prepared meat, using

Turkey breast, both wild and domestic, can be made into a mouth-watering muscle-meat jerky.

The Hi Mountain Turkey Hunter's Bourbon Blend Jerky Cure and Seasoning Mix creates a great-tasting, wild turkey jerky.

Debone the turkey breasts; remove skin, fat, and sinew; and then slice the meat into 1/4-inch-thick strips.

Weigh the sliced strips.

Lay the strips on a flat surface, and sprinkle with the proper amount of cure and seasoning mix according to the weight of the turkey strips.

the sprinkler bottle included in the package. Sprinkle the first side of meat with approximately half of the mixture. Next, turn the meat over, and sprinkle the remaining mixture on the meat. Put seasoned strips in a large mixing bowl, and tumble by hand until the mixture has been spread evenly on all sides of the meat. Stack the strips, pressed together tightly, in a nonmetallic container or sealable plastic bag. Refrigerate for at least 24 hours. Hi Mountain Jerky Cure and Seasoning is formulated to penetrate the meat at the rate of 1/4 inch per 24 hours; do not cure the meat any less than that. You're now ready to dry or dehydrate the jerky.

Sliced-Meat Turkey Jerky

2 lbs. wild turkey breast strips	2 tablespoons Morton Tender Quick
2 teaspoons black pepper	4 tablespoons brown sugar
1 teaspoon onion powder	1 teaspoon Liquid Smoke
1 teaspoon garlic powder	1 cup water or bourbon

You can also make up your own turkey-jerky cure and seasoning mix per the recipe above. Make sure cure, liquids, and seasonings are well mixed together. Pour the cure and seasoning mix over the strips, and then place the strips in a nonmetallic container. Mix well, making sure all surfaces of the meat are well coated. Cover and refrigerate for 24 hours. Remove, pat dry, and dehydrate or dry.

Drying

Small game jerky can be dried in a dehydrator following the manufacturer's instructions. It can also be dried in an oven set at 200°F.
The ground-meat jerky can be dried with any of the methods men-
tioned throughout this book. The turkey jerky strips can be placed on
wire racks or suspended in the oven. Make sure you have a pan below
the jerky to catch drippings. It's also a good idea to spray racks with
cooking oil to prevent sticking. Wild turkey meat tends to dry quicker
than red meat, so check the meat after about an hour's drying time.
Properly dried jerky should bend but not break. Meat made from fowl
must be heated to an internal temperature of 165°F to kill pathogens.
There also tends to be more oil, especially in domestic turkey meat.
This will bead up on the meat during the drying process. When the
jerky is done, pat dry any oil from the surface. After jerky has dried
completely, store in airtight containers in a dry, cool area. You can also
freeze and/or vacuum pack turkey jerky for longer storage.

Perfectly dehydrated turkey jerky is not only great looking but makes great eating as well.

④
Fish Jerky

DRIED FISH, IN one form or another, has been a staple food of mankind for thousands of years. Whether living near fresh or saltwater sources, people have regularly dried a variety of fish as a way of preserving it for future meals. Native Americans on both the East and West Coasts, for example, smoked and dried salmon for winter use. This smoked and dried meat would often keep until the next season. The eastern and western inhabitants also pounded the dried fish into a powder and traded it with the Native Americans of the Plains. Fish drying on racks in the traditional method is still seen in the coastal native villages of Alaska and Canada. Dried fish was also extremely important in Europe. In northern Europe, dried codfish is still a common food in many households.

The Native Americans dried fish such as salmon, split and hung it over wooden frameworks.

The Meat

It's recommended to use only lean-meat fish, including freshwater fish such as bass, brook trout, crappies, bluegills, walleyes, and perch. If using freshwater fish, it's important to use only fish that are fresh and completely free of parasites. Good choices in ocean fish include flounder, codfish, and even cuts of tuna that are free of fat. Oilier fish and those with a high fat content, such as snapper, mackerel, mullet, whitefish, carp, catfish, and pike, are not suitable for making into fish jerky. The high oil and fat content makes the jerky prone to spoilage. In many of these fish, it's almost impossible to trim off the fat, as the fat is evenly distributed throughout the flesh. If these fish are smoked, however, they can often be used with good results. In fact, smoked fish such as salmon and trout, although not technically a jerky because it is not dried, has long been a staple for many cultures and is a real delicacy.

Fish can deteriorate and decompose rapidly in heat as well as from the enzymes found in the flesh of fish. This is the main reason fish jerky should only be made of freshly caught and killed fish. The

Fish jerky or dried-salted fish has been a staple food source for many cultures. Fish with little oil or fat content, such as the largemouth bass shown here, are the best choices.

fish can be kept on ice a short period of time until you can make jerky from them. Fish should be free of slime, and the flesh should be firm with a slight fishy smell. Clean and fillet the fish as soon as you can to ensure freshness. Make sure the fillets are well cleaned in fresh, cold, running water to remove any blood. Cut the fillets into 1/4-inch-thick strips from 4 to 6 inches long. Do not make the strips any thicker or they will be hard to dry. If using small fish, such as bluegill, you may not need to cut the fillets into smaller strips.

To thoroughly kill any parasites present in fish, freeze the fish for at least 48 hours before curing or freeze the jerky after dehydrating.

Curing

Many types of fish used for jerky are commonly salt cured before dehydrating. This salt cure can be a dry or a liquid variety. Homemade cures can be used, or a number of cures are also available in retail stores. The following are two traditional methods. For a brine cure, use 1/4 cup of fine pickling or canning salt to 2 cups of water. If this volume doesn't cover the fish, prepare more brine following the same proportions of salt and water. This amount will brine 1 to 2 pounds of fish strips. Make sure the salt is thoroughly dissolved in the water, and then pour the brine over the strips. Use a glass container, cover, and place in the refrigerator for 48 hours. This salt curing not only draws out the moisture from the flesh, but it also aids in preservation and concentrates the amino acids.

You can also use a dry brine, allowing the dry salt to work into the meat and bring out the moisture. Place a thin layer of salt in a glass pan or dish. Apply a coating of salt to each fish strip and place in the pan. Layer the strips in the pan, making sure all strips are well coated with the salt. Cover the dish with plastic food wrap or a tight lid and place in a refrigerator for 48 hours.

In either case, remove the strips from the curing container, rinse under cold water, and pat dry. Now you're ready to dry or dehydrate. Simple salting and drying doesn't produce a very tasty fish jerky. To add taste, lightly coat the strips on both sides with soy sauce, Liquid Smoke, or Worcestershire sauce before drying.

Canning or pickling salt is used to cure the fish into jerky.

The fish must be filleted and then cut into strips about 1/4-inch thick and 5 inches long. Panfish fillets can often be used whole. Pat the strips dry.

Fish Seasoning Mix

....

1/2 cup brown sugar   1 teaspoon onion powder

1/2 cup salt   2 teaspoons white pepper

1 teaspoon garlic powder

....

You can also make a seasoning mix to add flavor to the fish strips. Above is a typical mix. Mix together and allow the mix to set overnight

Although salt alone will cure the meat, adding brown sugar, spices, and other flavorings adds to the taste.

Make sure salt, sugar, and spices are well mixed.

Spread the cure and seasoning mixture over the fish strips, making sure they are all well covered.

Place a layer of the cure and seasoning mixture in the bottom of a nonmetallic bowl or pan, and layer the treated strips in place, adding additional cure and seasoning mix as needed. Cover and allow to cure in a refrigerator overnight.

in an airtight container to blend the flavors. A pinch of powdered red pepper (or to suit) can be added to the mix if a hotter jerky is desired. Remove the strips from the original brine, rinse off the brine or salt, and pat dry. Coat the strips with the mix, and allow them to sit in the refrigerator for about 2 hours. Dry or dehydrate.

Cured Salmon

Fresh salmon	1 gallon cold water
1/2 cup white sugar	1 3/4 cups of Morton Tender Quick mix or
1/2 cup brown sugar	Morton Sugar Cure (Plain) mix

An excellent cured salmon recipe comes from Morton Salt. Clean and eviscerate salmon. Remove head, fins, tail, and 1/2 inch from each side along the belly incision. For salmon weighing less than 10 pounds, cut into 3-inch steaks. Split steaks along the backbone, leaving skin on if desired. If salmon is greater than 10 pounds, cut into 1 1/2-inch steaks.

Prepare 1 gallon of brine for each 5 pounds of salmon using proportions of water, sugar, brown sugar, and Morton Tender Quick or Morton Sugar Cure (Plain) mix as listed. Completely submerge the salmon in brine, using a ceramic plate or bowl to weigh down the fish and keep it submerged. Cure in refrigerator for 16 hours. Remove salmon, and rinse it in cool water. Pat dry and cook as desired.

Salmon may also be smoked in an electric smoker, following the manufacturer's instructions. The salmon must be heated to an internal temperature of 160°F and held at this temperature for at least 30 minutes. Refrigerate if not consumed immediately.

Drying

In the past, fish was simply dried outside. Typically, the Native Americans split the salmon carcasses and hung them over wooden frameworks in the shade but in an area with good air circulation and quite often with a smoky fire beneath. You can also dry fish outside, given the right circumstances, but make sure it is well covered to keep away insects. A smoky fire can be helpful not only in the drying but in keeping insects away. Fish can be easily dried in a dehydrator by following the manufacturer's instructions. Fish jerky can also be dried in an oven. Place on wire racks over cookie sheets to catch the drippings. Set the oven at 150°F and with the oven door slightly open. Dry for about an hour. Then, turn the strips over and dry for another hour. Test the jerky strips. The finished jerky should be like red-meat jerky. You should be able to bend it without breaking. It should not be crumbly or crunchy. When dried properly, there should be no moisture on the surface of the jerky sticks. For safety, the internal temperature must reach 160°F.

Storage

Properly cured and dried, fish jerky should last for a long time in sealed containers in a cool, dry place. For safety, you may prefer to freeze and/or vacuum pack the jerky strips. Storing the prepared jerky in the freezer will also take care of any parasites.

Smoked Salmon

Fresh salmon	1 gallon ice water
Hi Mountain Alaskan Salmon Brine Mix	

Although not an actual jerky, smoked salmon is easy to make with the Hi Mountain Alaskan Salmon Brine Mix. The brine mix is made with pure maple sugar. The package contains two bags of mix. Dissolve one packet in 1 gallon of ice water. The fish should be fresh and well chilled before curing. Immerse the fish in the brine solution, making sure it is well covered. Place it in a refrigerator for 24 hours.

Remove the fish from the brine, rinse it well with fresh, cold water, and pat it dry. Let the fish sit at room temperature for about 30 minutes and then smoke it. Smoking time can vary depending on the type of smoker, location, outside temperature, and so forth. The fish should be smoked until the internal temperature reaches 160°F. If you cannot get fish to the desired internal temperature with your smoker, place it in the oven to finish once the desired color is reached.

PART 3

Sausage

①
All About Sausage

The History of Sausage

Known as a staple food for almost a thousand years before Christ, sausage is one of mankind's oldest forms of processed foods. Homer's *The Odyssey* describes a form of sausage made from a goat stomach filled with fat and blood, roasted over an open fire. A Chinese sausage made of lamb and goat meat, called *lachang*, is recorded as early as 589 BC. The word *sausage* comes from the Latin word *salsus*, meaning salted or preserved; *salsus* was an extremely common food for the Romans. Sausages became so popular during the beginning of the Christian era that Roman emperor Constantine banned them.

National varieties of sausage originated in various regions and cities. Sausages are made with herbs, spices, and meats and include traditional ingredients that create special regional dishes. Through the Middle Ages, the English called it "sausage." In France, the term is *saucissons*, and in Germany, *wurst*. During the Middle Ages, sausage making became an art, with numerous commercial sausage makers scattered throughout Europe. In fact, these *wurstmachers*, as they were called in Germany, produced a number of distinctively flavored and spiced sausages that became known by the names of the cities or regions from which they originated and eventually became world famous.

The extremely popular frankfurter, or hot dog, came from Frankfurt, Germany. The *wiener*, however, is a product of Austria, the word meaning "Viennese" in German. Almost as popular, the luncheon meat bologna came from Bologna, Italy. Other famous sausages with city names include Arles, from France; Goteborg summer sausage, from Sweden; Genoa salami, from Italy; and braunschweiger, from Brunswick, Germany. With over 1,500 varieties of *wurst*, Germany has to be the sausage capital of the world. Sausage making was and is a serious business in Germany. During the fifteenth century, the Bratwurst Purity Law outlawed the use of rotten or wormy meat. Other famous German sausages include rindswurst, knockwurst, and bockwurst.

Sausages are also a very popular breakfast dish in the United Kingdom and Ireland, with well over 400 known recipes. A sausage,

Sausage, in all its varieties, is one of mankind's oldest and most important foods. Making your own sausage is not only an enjoyable hobby but also a valuable skill that can provide delicious food for your table.

dipped and fried, is very common in Britain, as is "saveloy," a pre-cooked sausage similar to (but larger than) the hot dog. Colored with brown dye, the sausage has a very distinct red color. A very popular snack food is the "pig in a blanket," a sausage cooked in a pastry. Another version is "toad in the hole," or sausage baked in Yorkshire pudding and served with onions and gravy. Square sausage is a popular breakfast food in Scotland. Seasoned mostly with pepper, it is formed into a block and cut into slices for cooking.

Scottish "black pudding" is similar to German and Polish blood sausages. The national sausage of Switzerland is *cervelat*, a cooked type of summer sausage. *Falukorv*, a traditional Swedish sausage, is made of pork and veal and contains potato flour and mild spices. It originated from the city of Falun. A fermented sausage called *sucuk* comes from Turkey and the neighboring Balkans.

It is made primarily from beef and is placed in an inedible casing that is removed before consuming. Some varieties may also contain sheep fat, chicken, water buffalo, or turkey meat.

Chorizo, a fresh sausage made of beef or pork salivary glands, is the most popular sausage of Mexico. It is often fairly dry, loose, and crumbly and used as a filling for torta sandwiches, tacos, and burritos. A moister and fresher version of chorizo is very popular in much of Latin America. A number of Philippine sausages include varieties of *longaniza* and chorizo. The traditional sausage of South Africa is called *boerewors*, or "farmer's sausage," and is made of game and beef with pork or lamb and usually contains fairly high amounts of fat. In Australia, English-style sausages called "snags" are popular, as is "devon," a pork sausage quite similar to bologna. New Zealand sausages are similar to those from England.

In Asia, popular sausages include Chinese *lap chong*, a dried-pork sausage that has some of the flavor and appearance of pepperoni; a ground-fish sausage from Japan; and *sundae*, a blood sausage and popular street food from Korea. *Saucisson* from France is a dried sausage, containing pork, wine, and/or spirits and salt. A number of regional varieties are made. Italian sausages are typically made of pork only

and usually contain fennel seeds, black pepper, and sometimes chilies or parsley. Swedish sausages are also typically made of mostly fine-ground pork and are lightly spiced. In Denmark, the popular "hot dog stand" serves *polser*, a very popular national dish. In Iceland, traditional sausages have been made of mutton and horse meat. Poland is well known for its variety of sausages, beginning with wild game meat from the royal hunting excursions. The sausages of Portugal, Spain, and Brazil, called *embutidos* or *enchidos*, are often highly spiced with peppers, paprika, garlic, rosemary, nutmeg, ginger, and thyme.

Early immigrants brought the tradition of sausage making to America; it eventually became a very important industry and remains so today. Native Americans had already learned to dry and cure meat such as venison, elk, and buffalo, and they also made a sort of sausage, combining spices, berries, and other ingredients with dried meat into a food product called pemmican.

Sausage is made from many different types of meat, using many different recipes. A common example of fresh sausage is breakfast patties.

The hot dog is America's favorite sausage, with the corn dog (a hot dog fried in cornmeal) also popular. Pork breakfast sausage in its many varieties is another favorite. America's melting-pot population enjoys a wide variety of sausage, including bratwurst, salami, Italian and Polish sausages, American-style bologna, liverwurst, head cheese, Cajun boudin, chorizo, and andouille.

Types of Sausages

Sausage is typically made from ground, minced, or emulsified meat. The meat may come from a single species, such as fresh pork sausage, or a combination of several different species, as in the case of hot dogs, bologna, summer sausages, and so forth. Almost any type of meat from domestic animals—mutton, goat, beef, pork, and fowl, such as chicken and turkey—to all types of wild game and fish can be made into sausage. Different types of sausages can utilize almost all parts

Another very popular type of sausage is dried or hard sausage, often called summer sausage because in some forms it does not require refrigeration. Wild game meat such as venison is often made into summer sausage, but beef and pork are also commonly used.

Cooked sausages are some of the most popular and include traditional hot dogs, bologna, braunschweiger, and other luncheon meats. Making your own is fun and a great way to use a variety of meats.

of the animal, which is the main reason for the popularity of sausage through the ages. As my grandmother used to say during hog-butchering days, "We use everything from the pig but the squeal."

Different types of sausages were developed in certain areas because of the types of meat that were found there. Sausages are seasoned in numerous ways, and the recipes are handed down through the generations, with the different recipes providing the variety of flavors. The meat is then cooked into patties or loaves or stuffed into casings. The casings can be either natural (made from animal intestines), artificial, collagen, synthetic, or sewn muslin.

Sausage is available in thousands of varieties worldwide, with different nationalities classifying them differently. In North America and other English-speaking countries, sausage is classified as three basic types: 1) fresh; 2) dry, summer, or hard sausage; and 3) cooked sausage. In other countries, the categories may be expanded to include fresh, fresh/smoked, dry, cooked, cooked/smoked, and so forth.

Other common forms are the specialty sausages, many of which, such as liver loaf and head cheese, are made into loaves and baked.

Fresh sausage is made from meat that is uncured and uncooked, and it must always be cooked before eating. Fresh sausage must also be consumed immediately or kept frozen. It won't keep any longer than fresh meat, even with refrigeration. In many parts of the world, the different forms of fresh sausage are often called breakfast sausage. The most common type of fresh sausage is pork, and even then, there are many different recipes, depending mostly on the flavorings used. Lean wild game meat such as venison and elk are also sometimes mixed with pork fat to create a fresh sausage. Fresh sausage may or may not be stuffed into casings. Stuffed fresh sausage includes bratwurst and other varieties, like Italian and Polish sausages.

Dry or *hard sausage* is sometimes called summer sausage because some types will keep during the summer or during warm weather without refrigeration. This was a very important means of preserving food before refrigeration and canning became available. These sausages are usually eaten cold. In some instances they are also fermented. Summer sausage is sometimes called "seminary" sausage because it is associated with monasteries. Dry sausage is made from cured meat that is either air-dried or commercially dried under controlled time, temperature, and humidity. Examples of dried sausage include salami, pepperoni, and the different varieties of summer sausage.

Cooked sausage is made of fresh meats that are already cooked, meaning they do not have to be cooked again before eating. These may or may not be smoked and include the very popular hot dogs and bologna and the many varieties of meat-and-spice mixtures, like pickle loaf. Although hot dogs do not have to be cooked, most people prefer to heat them before eating. Other examples include braunschweiger and liver sausage.

All types of sausages may or may not be cold-smoked for flavor. In addition to the three basic types, a wide range of specialty sausages exists, and the meats may be uncured or cured, chopped or comminuted. These are usually baked or cooked instead of smoked and formed into loaves. They are commonly served cold on salads and sandwiches and include such things as head cheese. Scrapple, another specialty sausage, is served as a breakfast meat.

Making your own sausage is an ancient skill that is fun to do. Using our guide, you can make up your own meat products with confidence, knowing exactly what's in them and how they are made.

②
Tools and Materials

YOU CAN MAKE sausage with little more than a mixing bowl and basic kitchen measuring tools, including a scale, measuring spoons, and spatulas. Simply purchase ground meat and add seasonings, then cook. But that really isn't sausage making. You'll probably want to be able to grind the meat, stuff the sausages, and cook or smoke them. All of these tasks require some specialized tools, in addition to the kitchen tools you may already have on hand.

You'll need a good, sharp knife for cutting meat. Having more than one knife definitely makes the different meat-cutting chores easier. The best way is to have a number of knives on hand in a variety of shapes for the many different tasks involved.

Although you can make sausage with nothing more than a few kitchen tools, serious sausage making requires a bit more equipment.

It's best to have a variety of knives available. From left to right: butcher knives, a hunting knife with gut hook for field-dressing, a thin-bladed boning knife, and two old-time skinning knives. Each type performs a specific task.

This is especially true if you do your own butchering as well. If you field-dress and skin big game, a skinning knife with a fairly short blade and a gut hook makes field-dressing and skinning easier. The blade shape should be rounded and with a drop point to make it easier to slice the skin from the muscle without cutting the skin. The same fairly short knife can also be used for cutting up the meat during the butchering process, but regular butcher knives, with longer blades, are best for this chore.

I inherited a number of old-time butcher knives, including some that were actually carried by trappers and explorers back in the 1800s, and I still enjoy using them. Butcher knives come in a wide variety of sizes and shapes, and again, the different types are used for different chores. Wide-bladed knives are best for slicing meat into the small chunks necessary to run them through a grinder for sausage. Thin-bladed, more flexible knives are best to use when deboning meat for jerky. I actually prefer the rounded-tip butcher knives to those with a sharp tip for cutting meat into smaller pieces. The upswept tip doesn't "catch" on meat as you slice it.

Photo courtesy of Bass Pro Shops

The RedHead Deluxe Butcher Knives Kit from Bass Pro Shops has a selection of knives suited to specific butchering chores, as well as a meat cleaver, meat saw, and sharpening steel.

You can also purchase butcher knife sets with a variety of knife shapes included. Some sets even include a sharpening steel as well. For instance, the RedHead Deluxe Butcher Knives Kit includes a paring knife, boning knife, butcher knife, meat cleaver, spring shears, square tube saw, honing steel, cutting board, butcher's apron, and six pairs of gloves, all in a carrying case.

It is important to always purchase and use quality knives. Not only are they easier to use, but they are also safer, because they will keep an edge longer and are easier to sharpen. When purchasing a new knife, you should ensure the blade has a sharp edge and resists dulling but is also easy to sharpen.

A variety of handle shapes and materials are used in knife construction. In quality knives, the handles will be most commonly made of hardwood or a synthetic material. The latter handles, sometimes made of soft-molded materials, are easy on the hands for long periods of use.

It's important to understand the types of steel used in knives. Knife blades are made of three different types of steel: carbon, stainless steel, and high-carbon stainless steel. Carbon was the original steel used in knife construction, and many old carbon-steel knives are still in use, including several in my collection. Carbon steel is relatively soft and sharpens very easily, even with nothing more than a handheld stone. It doesn't, however, hold an edge very well and must be continually resharpened. Carbon steel also rusts badly, even if the metal is dried after cleaning. One solution is to spray the metal with a light dusting of spray cooking oil before storage. Always wash the knife thoroughly before reusing to remove any residual oil and rust.

Pure stainless steel is the hardest of the three metals, but it is almost impossible to resharpen correctly at home. Stainless steel will, however, hold an edge nearly forever once sharpened, and it does not rust. The majority of the knife blades today are made from high-carbon stainless steel. This material provides the best of both worlds: It is a fairly easily sharpened blade that will hold an edge for a reasonable length of time, and it doesn't rust as badly as carbon steel.

Knife-blade edges are commonly ground into one of three shapes: flat-ground, hollow-ground, and taper-ground. Flat-ground blades have their edge ground evenly from the back of the blade to the front of the blade and from the heel to the point, then an edge ground. These blades are sturdy and easily sharpened. Hollow-ground edges have a portion of the blade just behind the edge thinned. This creates less drag, but it also creates a weak area in the blade. A better solution is a taper-ground knife. In this case, after the flat grind, an additional grind is made to thin out the blade but not to create the thinness of a hollow-ground edge. A taper-ground produces a knife with less drag, but with a stronger edge than a hollow-ground, and is found only on high-quality knives.

Sharpening

Keeping knife blades sharp is an extremely important facet of any type of meat preparation. Using the proper tools can make it easier to have sharp knives as needed. Many knife-sharpening devices are

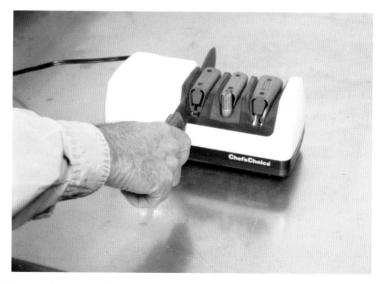

It's extremely important to keep knives sharp. A variety of sharpening devices are available. I've tested several Chef'sChoice power sharpener/hones for a number of years. They are extremely efficient, producing razor-sharp edges in seconds. Shown here is the Model 130.

The Chef'sChoice Diamond Hone AngleSelect Sharpener, Model 1520

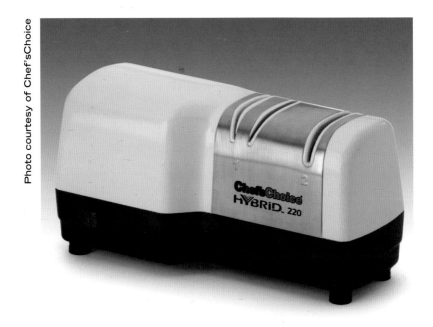

The Chef'sChoice Hybrid Sharpener 220

available, ranging from the simple but extremely effective butcher's steel to powered sharpeners. Knives that are in good shape other than being dull should never be sharpened using a powered grinder, as you stand a good chance of overheating the steel and losing the temper. Powered sharpening hones, however, can make the chore of sharpening a dull knife quick and easy.

I've tested the Chef'sChoice models for many years. The Chef'sChoice Professional Sharpening Station Model 130 is an excellent choice. It will sharpen both straight and serrated edges and has a 125-watt motor with three sharpening stages: a 40-degree presharpening stage, a 45-degree sharpening stage, and a third steeling stage for final sharpening. Springs guide the blade for precise sharpening.

Their Model 120 has similar features, except a stropping stage replaces the steeling stage for a Trizor triple-bevel edge.

The Chef'sChoice Diamond Hone AngleSelect Sharpener, Model 1520, is engineered to put a razor-sharp edge on all quality knives and can restore and re-create both a 20-degree edge for European- and American-style knives and a 15-degree edge for Asian-style knives.

Photo courtesy of Bass Pro Shops

One of the best hand-sharpening units I've tested is the Lansky Sharpening System. The system is available in three different kits that consist of hone holders color coded for easy identification of the exact sharpening angles desired.

An old-fashioned butcher's steel is the perfect tool to keep close at hand for regularly retouching the blades of knives as you use them.

The multibeveled, razor-sharp 15-degree edge on hunting knives reduces the amount of effort needed to cut or fillet, making it ideal for skinning and field-dressing.

A more economical sharpener is the Chef'sChoice Hybrid Sharpener. The Hybrid technology combines electric and manual sharpening and features two stages: an electric-powered stage for sharpening and a manual stage for honing. The Hybrid 220 couples the two stages with precise bevel-angle control to provide a super-sharp, arch-shaped edge that is stronger and more durable than the conventional V-shaped or hollow-ground edges.

Another system I've found extremely effective and easy to use is the Lansky Sharpening System. The kit contains finger-grooved hone holders that are color coded for easy identification of the exact sharpening angles desired.

All components are stored in a carrying case: knife clamp with angle selector, guide rods, extra attachment screws, oil, and a sharpening guide. Three kits are available: The Standard Kit has coarse, medium, and fine hones; the Deluxe Kit has extra-coarse, coarse, medium, fine, and ultra-fine hones; and the Diamond Kit has coarse,

Another sharpener to keep on the butcher table is the Lansky Hand Sharpener. A hand guard provides protection for your hand.

You'll need a nonporous work surface that is easy to clean; a secondhand, stainless-steel table like the one shown here is ideal, or you can use a large synthetic cutting board.

medium, and fine diamond hones. An old-fashioned butcher's steel is actually the quickest and easiest tool to use for keeping an edge on blades as you work. Merely draw the blade across the steel a couple of times as the blade becomes somewhat dull between cuts.

Another extremely simple and efficient sharpener for at-hand work is the Lansky Hand Sharpener. You'll want to keep either of these tools nearby for quick "touch-up" sharpening.

Work Table

These days, in addition to knives, a wide range of tools can make sausage making easier, faster, and safer. Next to a knife, the most important piece of equipment is a good, solid working table or surface that is easily cleaned. Your kitchen table or kitchen countertop will work, but it should not be wood or even the traditional butcher block; instead, it should be an easily cleaned, nonporous surface. Sausage comes out of a stuffer in a fairly long tube, and a large, flat work surface is necessary to catch and hold the sausage.

A number of years ago I purchased a pair of stainless-steel tables at a school auction. These are ideal, as they can be cleaned and sanitized. You can find tables like these quite often when restaurants hold going-out-of-business sales. In lieu of the ultimate table, a large, synthetic cutting board is the next-best option. Regardless of what is used, it must be easily cleaned and sanitized.

Grinders

If you're making only a small amount of sausage, a hand grinder may be your best choice. These are economical and easy to use. Of course, they are only as fast as you can hand crank them, and they do require a certain amount of muscle power.

Grinding meat for sausage requires a grinder. If using an old grinder, be sure it is well cleaned and the blades are sharpened before using. Small amounts of sausage can be run through a hand grinder. Hand grinders are available in clamp-on or bolt-on models.

Photo courtesy of Bass Pro Shops

I inherited an old hand grinder, and it had the usual problem of any well-used, well-worn grinder: The blade and plate were dull. In fact, the blade had nicks in it. I filed the blade smooth and used wet-or-dry sandpaper to finish smoothing it. I sharpened the plate by rubbing it across a flat hone, and soon, I was back in business. These old grinders are often found in antique shops and flea markets for a fairly cheap price, but be forewarned: You'll probably have to sharpen and refurbish them before actually using them. Most of the smaller units clamp on to a work surface. Larger models are made to be fastened down securely with bolts.

If you're purchasing new, Bass Pro Shops has a couple of clamp-on models and one bolt-on model that are excellent quality. Their heavy-duty bolt-on model grinds five to six pounds per minute and comes with 3/8-inch and 3/16-inch plates, grinding knife, and sausage-stuffing tube.

One method of fastening a grinder solidly is to screw the bolt-on models to a wooden board and then clamp the board to a work surface.

For many years I used a large antique powered grinder I had inherited from my granddad. It had a big flywheel and was powered

An electric grinder speeds up the work and makes the chore of grinding meat easier. Top choices, such as the LEM/Bass Pro grinders, are sturdy, with easy-to-clean parts of stainless steel, and are also easy to use. The No. 12 model shown grinds 360 pounds of meat per hour.

by a large electric motor with a belt. My granddad originally powered the grinder with his Model A, jacking up the rear wheels and running the grinder with a big belt. The throat or opening on this grinder is about three inches, and it didn't come with a meat pusher. It was (and is) dangerous—not only the grinder but also the spinning and flopping belt. Nonetheless, it was used by three generations of our family, without mishap.

The ultimate for sausage making is a modern-day powered grinder. They're available in a number of sizes, depending on the horsepower, which determines the amount of meat that can be ground in an hour. Smaller models will do the job, but with less power, they are slower. Top models can grind up to 720 pounds per hour. The LEM/ Bass Pro Electric Grinders, available from Bass Pro Shops, are some of the best on the market. They all feature a full two-year warranty and have heavy-duty construction, featuring stainless-steel, easily cleaned housing, grinder head, and auger; a permanently lubricated motor; built-in circuit breaker with roller bearings; all-metal gears; heavy-duty handle; and 110-volt power.

Standard accessories for all models include a large-capacity meat pan, meat stamper, stainless-steel grinding knife, stainless-steel stuffing plate, one stainless-steel fine (3/16 in.) plate, one stainless-steel coarse (3/8 in.) plate, and three stuffing tubes. Four models are available from Bass Pro Shops. The model I've tested is the No. 12, featuring a .75-horsepower motor, which is capable of grinding 360 pounds per hour.

Food Processors

Sausages such as frankfurters and bologna require the meat to be emulsified. Commercial sausage makers use big meat choppers for this task, with razor-sharp cutting blades to reduce the meat to extremely fine particles in minutes. Home sausage makers often do this chore by running meat through a grinder and a fine plate, up to

Some sausages, such as bologna and hot dogs, require extremely finely ground or emulsified meat. A home food processor is the best tool for this job.

three or four times. Another method is to use a kitchen food processor to fine chop the meat after it has been ground using a coarse plate. A little water must be added during the processing. A high-quality food processor is important for this task.

Mixing

Doing a thorough job of mixing the meats, seasonings, and other ingredients for a large batch of sausage can be very difficult without the help of a mixer. The LEM mixer from Bass Pro Shops will easily mix seventeen pounds at a time.

Photo courtesy of Bass Pro Shops

If you have a large amount of ground jerky meat to mix, the LEM 17-pound manual mixer can be a great helpmate.

Stuffers

Sausage stuffing is what makes sausage, and stuffers are available in several forms. The first is a simple plunger-type hand stuffer. These consist of a tube with a plunger and a stuffing tube on the end. They're fairly hard to operate and usually don't produce consistent sausages. If you like to make fresh sausage links with no casings, the LEM Jerky Cannon comes with a nozzle for stuffing sausage and resembles and operates much like a caulking gun.

The next step up is a lever-operated stuffer. This consists of a cast-iron L-shaped tube and lever to push a plunger down through the tube. As I discovered years ago, older models allowed a lot of sausage to escape back past the plunger because of back-pressure from the sausage. Newer models have a rubber or plastic gasket to help prevent this leakage problem around the plunger.

The stainless-steel stuffer is the easiest to fill and use.

Stuffers are available individually or in kits. The Bass Pro sausage stuffer kit shown includes a cast-iron plunger-stuffer; seasonings for fresh sausage, summer sausage, and bratwurst; and casings.

The Bass Pro LEM stuffer kit includes a 5 lb. vertical stainless-steel stuffer; seasoning for 60 lbs. of meat, including summer sausage, brats, and flavored sausage; and casings.

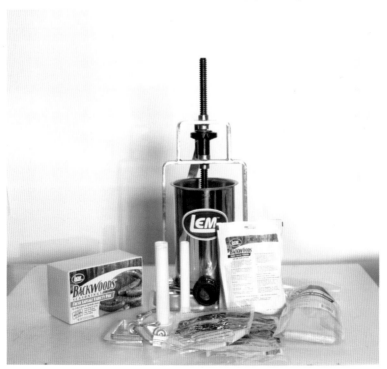

Two excellent models are available from Bass Pro Shops, including a 3-pound, cast-iron and 5-pound stainless-steel model. Both include 3/8-inch, 3/4-inch, and 1/8-inch stainless-steel stuffing tubes. The stainless-steel model is easier to clean.

To operate, fill the tube and push down on the lever. Vertical stuffers are the easiest to use, with the least amount of hassle, and produce more consistently stuffed sausages—especially when stuffing the smaller-diameter casings.

These are available in several sizes, including 5-pound, 15-pound, and 25-pound capacities. All are made of stainless steel, and the cylinder removes for easy cleaning. All come with different-size stuffing tubes. Bass Pro carries a complete line of vertical stuffers, including a commercial-grade model. Also from Bass Pro are two sausage-making

Fasten the stuffer to a wooden board that can be clamped to the work surface.

Grinders with stuffing tubes can make grinding and stuffing a one-step process.

kits complete with stuffers. The Sausage Kit includes a 3-pound cast-iron plunger-stuffer; seasonings for regular sausage, summer sausage, and bratwurst; and a bag of natural hog casings and fibrous casings.

A larger stuffing kit includes the LEM 5-pound vertical stainless-steel stuffer; enough Backwoods Seasoning to season sixty pounds of meat, including summer sausage, brats, and flavored sausage (hot, sweet Italian, and hot Italian); ten fibrous casings; two packages of natural hog casings; and a book on making sausage.

Regardless of the stuffer used, it is important it be solidly anchored to the work surface. Fasten the stuffer permanently either to a work surface or to a board clamped to the work surface.

A high-quality grinder such as the LEM/Bass Pro model shown also comes with stuffing tubes. You can grind and stuff many sausages at the same time. This is especially true with fresh sausages, when the meat is spiced before grinding.

Miscellaneous

The ends of casings of summer sausage and others can be fastened together with hog rings. For this task, you will need hog rings and hog ring pliers.

Many synthetic casings are fastened with hog rings. You will also need a hog-ringing tool.

To properly mix the ground meat and spices, you'll need a kitchen scale to weigh exact amounts of ground meat. Both 22-pound and 44-pound models are available from Bass Pro Shops. The 22-pound model comes with a stainless-steel tray and is sufficient to weigh most sausage batches.

We use our office postal scale, which digitally weighs up to 10 pounds and works well for small amounts of ground meats.

Other miscellaneous items needed for sausage making include latex gloves, measuring cups and spoons, glass bowls or other non-metallic containers, resealable plastic bags or containers, and cookie sheets or racks for drying sausages in an oven.

You will also need a thermometer. The best choice is a digital model with a temperature probe that stays in the meat or sausage when the oven door is closed. An alarm sounds when the desired internal temperature is reached.

A food-safe silicone spray can be used to protect and lubricate all parts of grinders and stuffers.

Although you can easily create your own sausage recipes, a wide range of premixes is available from a number of companies.

Cures for curing sausage are also necessary.

Sausages such as hot dogs, bologna, and others benefit from the addition of soy protein concentrate.

Photo courtesy of Bass Pro Shops

Most sausage recipes require specific amounts of meat. A kitchen-type (or similar) scale is necessary.

Food-safe latex gloves should be used for any field-dressing chores and are also a good safeguard when mixing sausages.

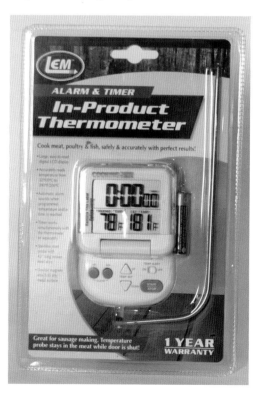

Most sausages require smoking or cooking. A digital thermometer with a remote sensor allows you to determine the internal temperature of the sausages while baking, boiling, or smoke/cooking.

Food-safe silicone spray can be used to lubricate and protect grinders and stuffers.

It's easy to make up your own sausage recipe, but you can also purchase premixed seasonings from Hi-Mountain, Bass Pro, and Bradley, as well as the LEM Backwoods sausage seasonings shown. The Sausage Maker, Inc., actually lists seventy-five various premixed seasonings, ranging from andouille sausage to boudin and venison salami.

The Bradley Smoker people also offer a line of cures for making sausages.

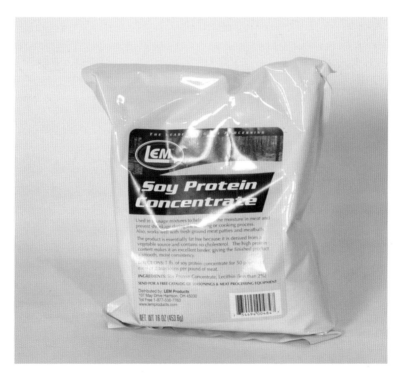

Some sausages also require powdered dextrose or corn syrup, as well as nonfat dry milk or soy protein concentrate.

Smokers

Many sausage products are smoked not only for flavor but in many instances for preservation as well. Any number of items may be used for smoking. Smoking actually consists of either cold- or hot-smoking. Cold-smoking is used to dry and flavor the meat; hot-smoking is used to cook the meat as well.

In the old days wooden smokehouses were a common farmstead building. These were used in late fall or winter to cold-smoke hams, bacon, and sausages. In cold climates some of the smoked meats were left hanging in the smokehouse until consumed. These buildings were usually large enough to allow for a small smoky fire to be built directly inside the structure, usually inside a small stone fire pit, or you could create an outside fire pit and pipe the smoke inside. My granddad had an old smokehouse, and even after he hadn't used it for years, it still held the smell of hickory smoke.

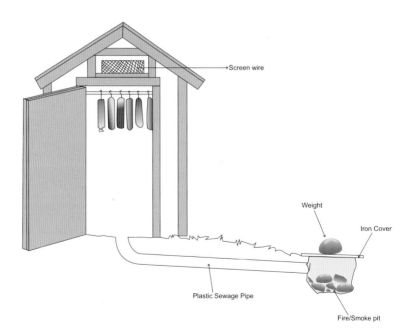

Wooden smokehouses were used to smoke sausages in years past. You can make your own.

Many years ago when I decided to try my hand at cold-smoking meats, I made a smoker from a discarded refrigerator, and it's still extremely effective. A refrigerator smoker can be "fired" in two ways: The first and original method is to cut a hole in the bottom or lower side of the refrigerator and install a metal pipe. The metal pipe runs a few feet away from the refrigerator to a fire pit. The pipe should be

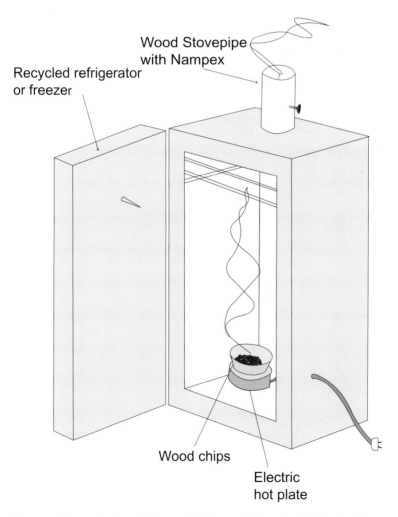

Wood Stovepipe with Nampex

Recycled refrigerator or freezer

Wood chips

Electric hot plate

Sausage smokers can also be made from an old refrigerator. This may be fueled by a fire pit or by using an electric hot plate.

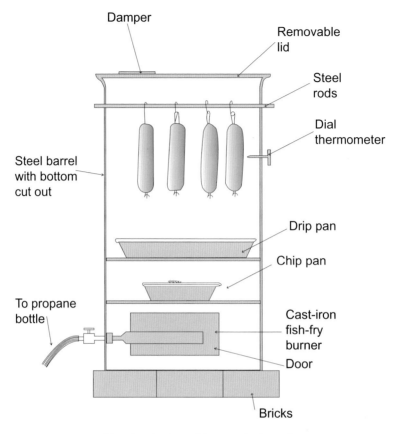

A barrel smoker is one of the easiest kinds to make.

buried (or covered with soil) for insulation. A smudge or smoky fire is built in the fire pit, and a sheet-metal cover is placed over the top of the pit to force the smoke into the pipe and refrigerator. A smoke opening must be cut in the top of the refrigerator, and a damper added. (A woodstove pipe with a damper can be used.)

The last step is to drill a hole near the top of the refrigerator and install a dial meat thermometer with silicone caulking around it. The only problem with both the refrigerator smoker and the old-fashioned smokehouse is that they both require a lot of attention to maintain the correct fire for smoking.

A number of years later, I added an electric hot plate to the bottom of my refrigerator smoker. Placing a pan of wood chips soaked in water

Electric smokers make smoking sausages easy and precise. The electric Bradley Smoker can heat up to 320°F for serious cooking, or the thermostat can be turned down to 200°F or lower for drying the meat.

on the hot plate made smoking easier, but I still had to monitor the thermometer and refill the chip pan.

You can also make a smoker from a large barrel, fueled by a propane bottle. This smoker can be used for both cold- and hot-smoking. Fire is provided by a cast-iron fish- or turkey-fryer burner with regulator. You will, of course, need to build a rack to hold the chip pan and

Bradley also has a cold-smoke adapter accessory that allows you to cold-smoke at lower temperatures.

The Bradley Smoker utilizes compressed wood-chip flavor bisquettes, available in a variety of woodsmoke flavors.

water/grease pan. A dial thermometer, a remote probe thermometer, and a damper on the top of the barrel smoker are also required.

A number of manufactured smokers are on the market these days, and many make smoking sausage much easier and more precise. These range from simple barbecue-style smokers utilizing charcoal to huge smokers capable of smoking a whole hog. The latter are dedicated smokers and can be used for cold-smoking and hot-smoking. Actually, the terms are a bit misleading. Pure cold-smoking doesn't cook the meat; it only adds smoke and dries the meat at temperatures under 130°F. Dedicated smokers, however, are primarily used to cook meat, although at a lower temperature and slower than in a direct-heat unit, such as a barbecue grill.

The Masterbuilt Electric Digital Smokehouse is a great choice for sausage makers. It not only has a digital thermometer with a meat probe, but the window also allows you to watch the smoking process.

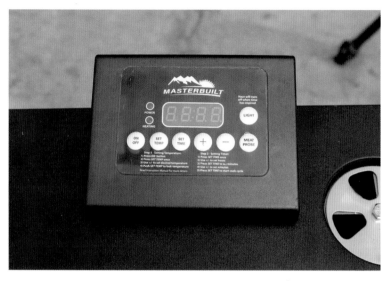

The Masterbuilt smoker makes it easy to regulate heat for cold-smoking and for cooking.

A wood-chip tray and water pan on the Masterbuilt smoker provides smoke and moisture.

An outside, removable grease tray on the Masterbuilt smoker makes it easy to collect and discard grease.

For instance, I smoke rib slabs at 225°F for almost an entire day but at much lower temperatures than you would find in a direct-heat grill. This is still hotter than cold-smoking.

One unit I've used for many years is the Good-One Smoker, and I've just about worn it out. These smokers are all quite expensive, but they can do double duty. Horizon Smokers, available from Bass Pro Shops, are an excellent choice, as are the Traeger smokers that utilize nontoxic wood pellets instead of charcoal. These smoke without an open flame.

The best choices for smoking sausage, however, are the electric smokers. One of the units I've used for years is the Bradley Smoker. Although mine is an older unit, the newer, digital models offer a number of advantages. Regardless, with either unit you can hot-smoke at up to 320°F or cold-smoke, slow-roast, or dehydrate down to 140°F merely by turning the thermostat to the desired temperature. A unique feature of the Bradley Smoker is that it burns Bradley flavor bisquettes, preformed wood-chip discs that self-load into the smoker. Precise burn time is twenty minutes per bisquette. The insulated smoking cabinet is easy to use at any time of year. Six removable

A vacuum packer can greatly extend the storage life of frozen sausages.

racks accommodate large loads and allow heat and smoke to circulate evenly. The Bradley has an easy, front-loading design.

The most versatile smoker I've ever tested is the Master-built Electric Digital Smokehouse. You can do anything from simple barbecuing to cooking sausages and smoking all types of meats. The digitally controlled electric smoker comes with a meat probe, an inside light, and a glass door. You can actually watch your meat cook and smoke. The push-button digital temperature and time control makes smokehouse cooking as easy as grilling. The thermostat-controlled temperature creates even, consistent smoking from 100° to 275°F. The built-in meat probe helps ensure perfectly cooked food every time and instantly reads the internal temperature of the sausage.

Wood chips are loaded through a convenient side door. The Masterbuilt smoker also features a drip pan and rear-mounted grease pan for easy cleanup. The removable water pan keeps food moist. The four removable, chrome-coated cooking racks have 720.5 square inches for large quantities of food. An air damper controls smoke.

Most sausages need to stay refrigerated or even frozen for longer-term storage. Vacuum-packing with a vacuum packer extends the storage life of your sausage. The better models have storage for the vacuum-bag rolls and a cutter built right into the unit. Vacuum packers are available online and from stores such as Bass Pro Shops, Walmart, and Target.

③

Sausage Basics

THE NUMBER OF steps involved in sausage making depends on the type of sausage being made. These steps include preparing and grinding the meat; seasoning/mixing; stuffing; and curing, smoking, and/or cooking the sausage.

Meat Preparation

Sausage is traditionally made from meat "scraps" and trimmings from butchering, utilizing many of the meat parts that are considered the less-desirable cuts. For instance, when butchering an animal, you'll have small trimmings from the shoulders, hams, steaks, and so forth. A typical hog will provide about fifteen pounds of trimmings from these cuts. Other commonly used meats include the neck, feet, heart, liver, tongue, and head.

Sausages have traditionally been made out of the trimmings from butchering animals. Butchering day at the Burch family farm was usually in January, so the meat would cool down properly and stay cool during the time required to butcher and make the various food products, including sausage.

Good sausage requires some fat; the most common ratio is 75 percent lean meat to 25 percent fat, although the percentage can vary according to taste. In many instances you will need to trim some fat from the lean meat.

If you are making any quantity of sausage, you'll need additional meat. Sausage can be made from the entire animal, with whole-hog fresh sausage being especially prized because the entire carcass, including the loins, hams, and shoulders, is utilized. Good sausage normally contains about 75 percent lean meat and 25 percent fat, although this will vary somewhat depending on the recipe and personal preference. Some people prefer a two-third lean and one-third fat combination, while others prefer an 80 to 20 percent mix.

When making sausage of lean meat—such as beef and especially venison—pork fat is usually added. In some countries mutton is used for the fat. Fat allows for easier frying of fresh sausage and helps bind the lean meat in other types of sausage. Fat is especially important when using any extra-dry meat, such as venison. Too much fat, however, can also be a problem. Excess suet should be trimmed from beef, and because of the potentially "gamey" flavor, all fat should be removed from venison. It's also important to keep pork fat levels low in all pork sausages.

The only time I remember my grandparents disagreeing was during hog-butchering days. The Burch family fresh pork sausage was typically made mostly from the shoulders of the hog, with some trimmings from the sides added. The hams were sugar cured. Grandma wanted more of the sides put into the sausage, and Granddad wanted more of the sides to slice and fry and make into bacon. If you aren't butchering your own animals but purchasing meat for sausage, many recipes call for "pork butt" or pork shoulders, often called "Boston butt."

A typical bone-in pork shoulder with the skin on will weigh from fifteen to twenty or more pounds. Skinning and deboning will usually result in from thirteen to eighteen or nineteen pounds of meat. A boneless, skinless pork shoulder will provide just about the correct amount of lean meat and fat for a pure pork sausage (i.e., fresh breakfast sausage). However, you may find that you prefer more or less fat content. Regardless, it's important to use only clean, fresh meat ingredients. In most instances when you purchase a pork butt or pork shoulder, it will be bone-in.

If you purchase meat for sausage, a pork shoulder—called pork "butt" or sometimes "Boston butt"—is a good choice. This cut usually has about the right amount of fat and lean.

You may, however, need to trim some of the outside fat away, and this cut will have a bone. Boning out, however, is not a problem. Make sure you cut in and around the protruding "blade bone" to get all the meat.

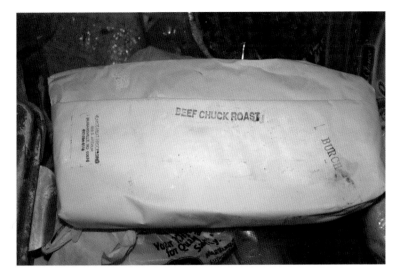

A chuck roast or beef plate is the best choice for beef sausage.

Venison is a very popular sausage meat. Make sure you trim away all fat and sinew, as the fat can cause the meat to taste "gamey."

Beef trimmings are also traditionally used for sausage making by home butchers. If purchasing beef for making into sausage, economical cuts include chuck roasts and beef plates. A very common meat for deer hunters to use in sausage making is venison. Again, if one is butchering at home, trimmings are used, but successful deer hunters often make "whole-deer" summer sausage, a favorite of ours. With mutton and lamb, the whole carcass is often used for sausage. Turkey and chicken can also be used to make sausage or are often added to other meats to make a less-red-meat type of sausage. Turkey legs are an excellent poultry sausage meat.

Regardless of the meat you use, debone and trim all gristle, sinew, excess suet or fat, and blood spots from the meat. Cut the meat and fat into 1-inch cubes, but keep them separate. Weigh the fat and meat separately and divide into the mixture of fat and meat desired. Then measure all the other ingredients for the sausage recipe being used.

Grinding

You can have the butcher grind the meat for you, and this is a good choice if you simply want to try your hand at sausage making before purchasing a lot of equipment. But having the meat ground can be costly, and you'll probably want a grinder so you can control the quality of meat and ratio of fat to meat. Meat grinding can be done with a hand grinder or with a powered grinder. The latter, of course, makes the chore easier and faster, and if you prepare any amount of sausage, you'll eventually want to invest in a good electric grinder. Most electric grinders can also be equipped with stuffing tubes, making that chore easier as well.

Grinding meat for sausage is fairly easy, but it's a good idea to follow a few basic rules. To maintain the desired ratio of fat and meat, you should weigh the meat and fat and grind together or weigh them into two separate piles or containers and grind separately. Lean meat grinds easier than fat, but in either case, well-chilled meat will grind

Serious sausage making requires grinding the meat.

If using different types of meat in a sausage, keep the meats separate. Weigh the trimmed meat and/or fat pieces separately.

Cut the trimmed meat into 1-inch chunks for easier grinding.

the easiest. Meat should be 40°F or colder. Make sure there are no bones in the meat to stop an electric grinder or damage the worm gear and grinding plate and blade. A bone will definitely stop a hand grinder and is less likely to damage the grinder.

Grind the meat, making sure you keep the meat cold (40°F or colder). Grinders usually come with two grinding plates with different-size holes, fine and coarse. Different sausage recipes call for using different plates.

Some sausages require finely ground meats. In this case the ground meat should be placed in a flat pan and partially frozen to allow it to firm up.

Feed the meat slowly into the throat of the grinder. Use the meat stomper to push meat into the throat of the grinder head. Do not force the meat, and never use your fingers to push meat into the head. Used properly, today's grinders are very safe, especially compared to older versions, like my granddad's big grinder with a big, wide-open throat. The family joke—"Don't get your tie in there"—was really a reminder to all users to be extremely careful. Make sure you keep the knife and plate sharp and clean. A dull knife or clogged plate tends to crush out the meat juices, retaining more sinew and gristle, slowing or even stopping the grinding process, and reducing the quality of the sausage.

Grind the meat through the coarse (3/16-inch or 1/4-inch) plate first. Many types of sausage, including fresh, are made from this coarse-ground meat. Actually, any meat product used for frying, such as breakfast sausage, should be coarse ground, or it can become too dry to fry properly.

You may prefer to grind through a finer plate. Some sausages, like bologna and wieners, must be made of finer-ground meat. In this case, the meat is re-ground through the finer-grind (1/8-inch) plate.

Cut the partially frozen meat into 1-inch chunks and re-grind the chunks.

In some cases you may wish to grind through the fine plate as many as three times.

It's important to rechill the meat for each grind, as the grinding process actually heats up the meat as it goes through the machine. The

Repeat if you'd like an even finer grind.

meat should be thoroughly chilled after each grind. The best tactic is to store the coarse-ground meat mixture in a covered container and refrigerate overnight to allow the meat to firm up. Another technique to make the grinding even easier is to soft- or partial-freeze the mixture to about 25° F and cut into 1-inch cubes. Run these cubes through the grinder while they are still frozen. Fat grinds best using the same technique. If you want an even finer meat consistency, again, freeze the meat to 25°F, cut into chunks, and run through a food processor in small amounts.

With most sausages, the fat is ground last and added to the mix. This is particularly so with dry, hard sausages such as salami, which show the fat particles in the sausage as part of the recipe.

To change to the fine-grind plate, turn off the motor and unplug the grinder. Then, remove the coarse plate and clean the head of the sinew, fat, and gristle that has accumulated during the first grind. There is no way to remove all of this material from the meat scraps; some will always remain and be caught in the grinder. Reassemble the unit with the fine plate, plug in the grinder, and re-grind the meat.

The ground meat can also be partially frozen, cut into 1-inch chunks, and run through a food processor for emulsifying.

Take the grinder apart and thoroughly clean with hot, soapy water and a bleach/water solution. Spray all parts with a food-grade silicone to prevent rust and to keep your grinder in good condition.

If at any time during the first coarse-grind, or during any of the finer grinds, the meat mashes instead of coming through the plate in "strings," first unplug the grinder; next, remove all meat from the grinder and plate, as well as any sinew that may have collected; reassemble and tighten the grinder ring, tighter than it was previously; and then continue grinding the meat. When you're through grinding, run some saltine crackers through the grinder to help clean out the last bit of meat. (It doesn't hurt if some of the crackers are mixed in with the sausage.)

Once you're through grinding, unplug the grinder and disassemble the grinder head. Wash all parts in hot, soapy water, followed by a bleach/water solution. Rinse thoroughly in hot water. Allow the parts to dry completely. Spray the grinder parts with food-grade silicone to prevent rust and to keep your grinder in like-new condition while stored.

Curing and Seasoning

All sausages require seasonings, and some also include a cure. The seasonings and cure can be applied to the meat before or after grinding. In the case of fresh or breakfast sausage, the cure and seasonings are sprinkled on the meat chunks before grinding, then the meat, cure, and seasonings are well tossed and stirred before the meat is ground.

Another technique is to mix together the ground meat and ingredients. This is necessary if you utilize previously ground (and often frozen) meat. This method is especially effective if, for instance, you butcher a deer, grind the meat and then freeze it, and then later accumulate pork or another deer or two to be used with it (or if you wait until you have collected enough other wild game to warrant the time to grind and clean up the grinder). In this case, it's a good idea to package the ground meat in 1-pound resealable plastic bags or, even better, in vacuum-packed bags and then freeze. You can then get out the varieties of meat in the proportions needed for a specific recipe.

Sausages require seasonings, and a wide variety of recipes using different seasonings is available. The simplest method of seasoning sausage is to purchase a prepared seasoning mix.

You can easily make your own sausage seasoning mixes to suit your family's tastes.

In some cases, such as when making certain kinds of fresh sausages, the seasonings are sprinkled on the cut-up meat chunks before grinding.

The seasonings may also be mixed with the ground meat.

Sausage Ingredients

Besides meat, sausage also requires other ingredients, including salt and spices. Salt is necessary for curing and preservation, and the spices enhance the taste of the sausage. You can purchase the ingredients separately and make up your own blend following any number of recipes, or you can purchase premixed cures and seasonings, ready to use and with "guaranteed" results. The latter are available from a wide range of butcher-supply companies, many on the Internet; you can also get small batches from a local butcher or butcher-supply house.

Nitrates and nitrites have traditionally been used to cure meats, with potassium nitrate (saltpeter) and Prague powder supplying the nitrates and nitrites. According to the National Center for Home Food Preservation:

> Salt is an essential ingredient in sausage. Salt is necessary for flavor, aids in preserving the sausage, and extracts the 'soluble' meat protein at the surface of the meat particles. This film of

Meat cures are also added to some sausages. Use only the amount called for in the recipe.

Salt is also a necessary ingredient of sausage, although you can vary the amount to some degree to suit your taste.

protein is responsible for binding the sausages together when the sausage is heated and the protein coagulates. Most sausages contain 2 to 3 percent salt. Salt levels can be adjusted to your taste. Iodized or table salt is not the best choice; instead, use a coarse or canning salt. Other salt choices include kosher or sea salt.

A wide range of spices provides the flavors of the various types of sausages. Most of these spices are available at your local grocery, but some may have to be purchased from butcher-supply houses. Fresh spices are the best, as they add the most taste. Many spices can lose their flavor over time, especially if kept longer than six months at room temperature. The best storage for spices and seasonings is below 55°F and/or in airtight containers. Grinding your own fresh spices is the best method, as it provides the freshest flavor.

According to the National Center for Home Food Preservation:

Nitrates and nitrites are curing agents required to achieve the characteristic flavor, color and stability of cured meat. Nitrates and nitrites are converted to nitric oxide by microorganisms and combine with the meat pigment myoglobin to give the cured meat color. However, more importantly, nitrite provides protection against the growth of botulism-producing organisms, acts to retard rancidity, and stabilizes the flavor of cured meat.

Extreme caution must be exercised in adding nitrate or nitrite to meat, since too much of either of these ingredients can be toxic to humans. In using these materials, never use more than called for in the recipe. A little is enough. Federal regulations permit a maximum addition of 2.75 ounces of sodium or potassium nitrate per 100 pounds of chopped meat and 0.25 ounce sodium or potassium nitrite per 100 pounds of chopped meat. Potassium nitrate (saltpeter) was the salt historically used for curing. However, sodium nitrite alone, or in combination with nitrate, has largely replaced the straight nitrate cure.

Since these small quantities are difficult to weigh out on most available scales, it is strongly recommended that a commercial premixed cure be used when nitrate or nitrite is called for in the recipe. The premixes have been diluted with salt so that the small quantities which must be added can more easily be weighed. This reduces the possibility of serious error in handling pure nitrate or nitrite. Several premixes are available. Many local stores stock Morton Tender Quick Mix and other brands of premix cure. Use this premix as the salt in the recipe and it will supply the needed amount of nitrite simply and safely.

Much controversy has surrounded the use of nitrite in recent years. However, this has been settled and all sausage products produced using nitrite have been shown to be free of the known carcinogens. Remember, meats processed without nitrite are more susceptible to bacterial spoilage and flavor changes, and probably should be frozen until used.

Note that many recipes call for holding the meat overnight to cure. This is required to allow the bacteria to convert the nitrite to nitric oxide. The addition of a reducing agent such as ascorbic acid (Vitamin C) speeds the curing reaction and eliminates the holding time. Another reducing agent, sodium erythorbate (isoascorbic acid), may also be used. Meat inspection regulations allow the use of 7/8 ounce per 100 pounds of meat.

If you're a home sausage maker, however, you may prefer to simply allow for the curing time.

Morton Salt has a family of curing salts especially designed for curing meat in the home. The Morton Tender Quick Mix is a fast cure product that has been developed as a cure for meat, poultry, game, salmon, shad, and sablefish. It is a combination of high-grade salt and other quality curing ingredients that can be used for both dry and sweet pickle curing. Morton Tender Quick Mix contains salt, the main preserving agent; sugar; both sodium nitrate and sodium nitrite, curing agents that also contribute to the development of color and flavor; and propylene glycol, to keep the mixture uniform. It can be used interchangeably with Morton Sugar Cure (Plain) Mix. It is not a meat tenderizer.

Morton Sugar Cure (Plain) Mix is formulated for dry or sweet pickle curing of meat, poultry, game, salmon, shad, and sablefish. It contains salt, propylene glycol, sodium nitrate and sodium nitrite, a blend of natural spices, and dextrose (corn sugar). The company advises caution, noting, "These curing salts are designed to be used at the rate specified in the formulation or recipe. They should not be used at higher levels, as results will be inconsistent, cured meats will be too salty, and the finished products may be unsatisfactory. The curing salts should only be used in meat, poultry, game, salmon, shad, and sablefish. Curing salts should not be substituted for regular salt in other food recipes. Always keep meat refrigerated (36° to 40°F) while curing."

The spices used in Morton mixes are packaged separately from the other ingredients. This is to prevent any chemical change that may occur when certain spices and the curing agents are in contact with each other for an extended period of time. If you do not need an entire package of Morton Sugar Cure (Plain) Mix for a particular recipe or must make more than one application, prepare a smaller amount by blending one and a quarter teaspoons of the accompanying spice mix with one cup of unspiced Morton Sugar Cure (Plain) Mix. If any portion of the complete mix with spice is not used within a few days, it should be discarded. It is not necessary to mix the spices with the cure mix if spices are not desired. The Morton Sugar Cure (Plain) mixes contain the curing agents and may be used alone.

Another product, Morton Sausage and Meat Loaf Seasoning Mix, is not a curing salt. It is a blend of spices and salt that imparts a delicious flavor to many foods. The seasoning mix can be added to sausage, poultry dressing, meat loaf, and casserole dishes, or it can be rubbed on pork, beef, lamb, and poultry before cooking.

Soy protein concentrate is often added to sausages such as hot dogs and bologna as a binder for the finely ground meats.

In addition to curing and seasoning agents, many sausage recipes call for additional ingredients, such as starter or fermenting cultures (lactic bacteria) for fermented sausages. Other ingredients include extenders and binders. Commercial sausage makers use extenders to reduce the cost of the sausage, but extenders are not commonly used by home sausage makers. Binders improve the flavor, help retain the natural juiciness, prevent shrinking during smoking or cooking, and provide a finished sausage with a smooth, moist consistency. Nonfat dry milk, cereal flours, and soy protein products are the most commonly used extenders. Soy protein has no taste, contains no cholesterol, and is fat free because it is derived from a vegetable source.

Water is also added to most sausage formulations, and the amount used depends on the type of sausage. Water primarily replaces the moisture lost during smoking and cooking. Approximately 10 percent water is added for moist types of cooked sausage. Fresh sausage is usually made with very small amounts of water, usually less than 3 percent. This only aids in stuffing, mixing, and processing. Dried sausages, such as summer sausage and pepperoni, require no additional water.

Mixing

As you can guess, it's extremely important to thoroughly mix the different meats, cures, and seasonings together. Do not use your bare hands to mix the sausage ingredients; instead, always wear a pair of new, food-safe gloves. If you add cures and seasonings prior to grinding, the ingredients will be pretty well mixed when they come through the grinder. If you add the cures and seasonings after grinding, first mix the different meats (i.e., lean venison and fatty pork) together. Then, add the curing/seasoning ingredients, sprinkling in a little at a time; turn and mix thoroughly; add more and mix again. Overall mixing time, however, should be fairly short, about three minutes.

Keep the meat cool, or rechill if necessary. A large plastic tub is the best choice for mixing. These are available as meat tubs, or you can

It's important to make sure the seasonings and cures are well mixed with the meat. But do not overwork, and make sure to keep the meat cold. Wear clean, food-safe gloves when mixing meat.

substitute large plastic household tubs. Some recipes and techniques call for allowing the cure/seasoning sausage mix to set overnight in the refrigerator; however, this will cause the mixture to set up stiff. If the sausage is to be stuffed, it will be easier to do so if you stuff it as soon as it is ground and mixed with the other ingredients. Salt, as well as other ingredients—like soy protein and nonfat dry milk—all cause the mix to harden.

Fresh sausage is best produced by simply sprinkling on the seasonings, grinding, and stuffing (if it is to be stuffed). An alternative is to simply freeze the fresh sausage for future use, especially if it is to be cooked as patties. This can be done in two ways: 1) Make up the patties, place waxed paper between them, then place them in a resealable plastic freezer bag. All you need to do is pull out the number of patties you want to cook and add them to the frying pan. 2) Or you can simply place the fresh sausage in resealable bags for freezing.

An even better tactic is to vacuum-pack the sausage, as it will keep longer. It's a good idea to try to keep packages consistent in size.

Fresh sausage to be made into patties can be frozen for future use. We press sausage into plastic bowls to achieve a consistent amount of meat, and then place in plastic bags and freeze.

We simply press the sausage mix into a plastic bowl, usually a recycled whipped topping bowl, which makes up about one pound to one and a half pounds, and then place it into the bag, either freezer or vacuum-pack. Sausage may also be formed into loaves or rolls and baked. Regardless, unless the sausage is to be smoked, or cooked later, you may wish to fry up a bit to taste your mix before freezing. This is where you can really enjoy your work.

Sausage Casings

Many sausages are stuffed into casings to hold the ground meat in shape while it is cooked or smoked and also to determine the size and shape of the different sausages. Casings must be strong enough to hold the meat, yet pliable enough to allow for contraction and expansion during stuffing and cooking or smoking. Casings are available in several forms—natural, edible synthetic, synthetic/fibrous, cloth, and plastic.

Natural casings, the traditional form, are edible, are permeable to water vapor and smoke, and have been used since the earliest days of sausage making. Natural casings are made from the cleaned intestines

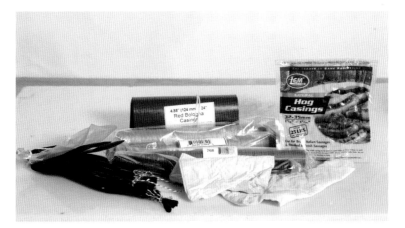

Many sausages are stuffed into casings, which are available in several forms and sizes.

Natural casings from sheep, hogs, and cattle, such as those used with these brats, are traditional and available in a number of shapes.

of animals, including sheep, hog, and beef casings. The inner lining is removed and the outer submucosa layer of collagen utilized. These casings are available in a wide range of sizes, depending on the source. Sheep casings are about 3/4 in. in diameter and are commonly used in fresh breakfast sausage to create links or for frankfurters. Hog casings come in several different sizes and are the most commonly used. Hog casings are usually about 1 inch in diameter but can be obtained in larger sizes and are used for Italian sausage, pepperoni, and large frankfurters and hot dogs. Other less commonly used hog casings include hog bungs and hog bladders, used for liver sausage and head cheese.

Beef casings are used for cooked and smoked sausages, such as salami, bologna, Polish sausage, and others. The most commonly used beef casings are beef rounds and middles. Other beef casing products include beef bungs, used for large bologna; beef bladders, used for round Mortadella; and beef rounds, used for ring bologna, ring liver sausage, and others. Usually recipes will recommend the size of casings to be used. Some beef casings are rather tough and are commonly peeled away before the product is consumed.

Most of the common natural casings, except bungs, bladders, and so forth, are sold in lengths, usually of several feet, called "hanks," or they are sold in bulk, in yards. A hank (or small container), around 60 feet of hog casings, will normally stuff from forty to fifty pounds of sausage.

Natural casings are packed in salt and must be prepared at least two to three hours before use. Remove the casings from the packing

Natural casings are packed in salt and must first be prepared for use by soaking in cold water.

Open one end of the casing and flush water through the casing. Keep the casings in cold water until use.

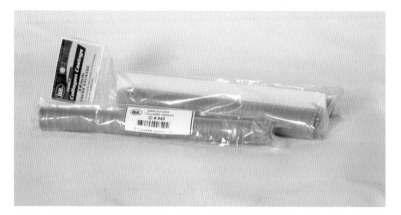

Edible synthetic or collagen casings are also available in several sizes and do not require as much preparation as natural casings.

and cut into 3- or 4-foot lengths for ease of handling. Rinse in cold water to remove the salt. Using two fingers inserted into one end of the casing, open the casing end and hold under running water to flush the casing. Pinch off the ends and shake the water around in the casing a bit. Check for any leaks indicating breaks. Next, place the casing in a bowl of water and soak for two to three hours. Changing the water after the first hour also helps.

If you must hold the casings for a short time, add a bit of vinegar to the water. Make sure the casings are wet before placing on the stuffing horn. Tough casings are usually caused by not allowing them to soak thoroughly before use. Leftover casings should be drained of any water and repacked in salt in the original container. Stored in a refrigerator, they will keep for about a year.

Edible synthetic casings are made from collagen, derived from the protein found inside the hides of pigs or cattle. This ground material, processed into a bread-dough-like mass, is extruded through a die of the desired diameter. In commercial sausage making, the collagen and meat blend is coextruded and then the outside of the sausage is coated with vinegar to cause it to set up. Collagen casings have become quite common, mostly because they allow for more economical sausage production. They are available in the same basic sizes as natural casings but do not need as much prestuffing preparation.

Synthetic fibrous casings are used for the larger sausages, and are peeled or cut off before consuming the sausage.

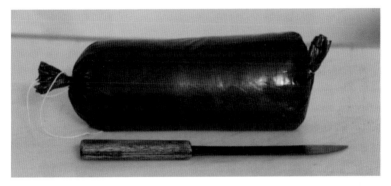

Nonedible, impermeable plastic casings are available for making skinless hot dogs or bologna.

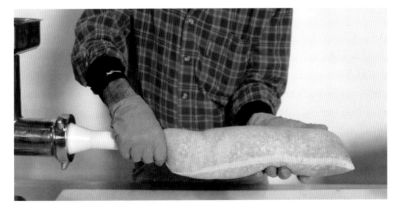

Cloth or muslin casings are also used for making many sausages.

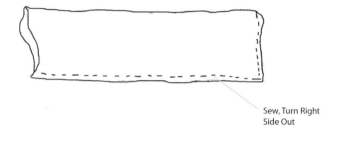

Sew, Turn Right
Side Out

Muslin casings can easily be sewn at home.

These casings are also permeable to both water vapor and smoke. Both natural and collagen casings tend to have an odor when you first open the packet they are contained in; however, this odor dissipates fairly quickly and is nothing to be concerned about.

Synthetic fibrous casings are also often used for sausages that will be sliced, such as summer sausage, bologna, and so forth. In smaller sizes, the synthetic fibrous casings are used to produce "skinless" wieners and franks. After cooking the casings are skinned off. These fibrous casings are also available for larger sausages and are commonly used for salami and summer sausage. They are not edible but packaged dry and ready for use. They often come with one end tied closed. If you want a smoked flavor, synthetic casings are available with smoke flavor added to the inside of the casings. The casings are available in different colors, and you can also get fibrous casings with deer heads stamped on them, identifying deer or wild game sausage. Fibrous casings are available in sizes from 3/4 inch to 6 inches in diameter and can be used for a wide variety of sausages.

Impermeable plastic casings are also available. These casings are not edible and must be peeled off before eating. They are most commonly used for hot dogs or bologna.

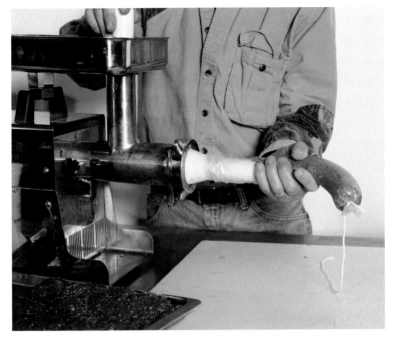

You really begin to feel like a sausage maker when you're stuffing sausage, which can be done in several ways. One of the easiest ways is to stuff at the same time you grind, using a grinder with a stuffing attachment.

Cloth or muslin casings have traditionally been used as well. Our family has used these homemade casings for stuffing fresh breakfast sausage for many years. The sausage is hand-stuffed into the casings and the rolls and then frozen. To cook the sausage rolls, slice them to create patties, remove the muslin, and then fry the shaped patties.

Muslin casings are sewn into the sizes needed. For breakfast sausage we create sewn casings that are 2 inches in diameter and about 16 inches long. To make the casings, sew 5" x 18" strips of folded-over muslin into "tubes." Muslin casings should be soaked in water and then wrung out before using, or they will dry the sausage as it is stuffed into them. I also use these casings for making venison summer sausage.

Stuffing Sausages

Stuffing is where the fun comes in—when you feel like you're really making sausage. Stuffing may be done by hand, with handheld stuffers, with stuffing tubes attached to meat grinders, or with some food processors. My mom hand-stuffed our breakfast sausage into muslin tubes, as we didn't have a sausage stuffer. I remember it was a fairly long and difficult process. When I began to make summer sausage from deer meat many years ago, I used a very primitive hand stuffer. It consisted of a tube with a stuffing tube on one end and a plunger pushed in by hand. It also was a lot of work. Then I acquired a 3-pound lever-operated stuffer from The Sausage Maker, Inc., and it made stuffing much easier. A better choice is a gear-driven, hand-cranked stuffer, such as the LEM model shown from Bass Pro Shops, and the ultimate is a quality grinder with stuffing attachment.

You can also stuff using a plunger-operated stuffer. Make sure the stuffer is solidly fastened to a work surface. I fasten the stuffer to a board that is then clamped to the work surface.

Make sure you have everything on hand before you begin stuffing. Meat should be chilled to below 40°F or slightly frozen before stuffing. The quicker you can stuff, the better. The first step is to choose the casing size and type desired for the specific sausage, then match the stuffing tube to the sausage size. Make sure the stuffer is solidly affixed to a large, cleanable work surface, as the cased sausage will be extruded out onto this surface. One method I've used is to screw the stuffer to a 3/4 in. wooden board and clamp the board with the stuffer to a table or countertop with C-clamps.

Sausage can be ground and then stuffed, or stuffed at the same time it is ground, depending on the type of sausage and the equipment you have available. Regardless of the method used, sausage ingredients should be ground as soon as you get them mixed. Some older recipes call for letting the mix sit for a certain length of time, usually overnight, so the ingredients can blend together. All this does is allow the salt in the cure and seasonings to set up the sausage, making it almost impossible to stuff. The ingredients will blend just as well

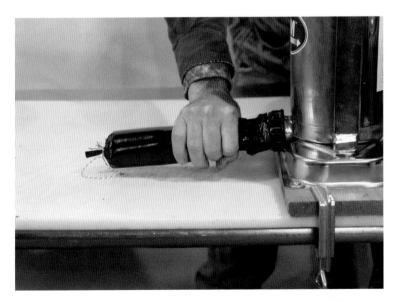

A gear-driven, hand-cranked stuffer can process a lot of sausage in a short time. Make sure you match the stuffing tubes to the size of casings being used and the type of sausage being made.

in the casings. Adding one ounce of water per pound of meat during the mixing process will also help make stuffing easier. This allows the meat to flow easier from the stuffer, filling the casings more uniformly. Water will not weaken the flavor of your sausage.

If you have purchased a new stuffer, read the instructions for your particular stuffer. If you've purchased or inherited an older-model hand stuffer, the steps are basically the same, regardless of the model or type.

Regardless of what type of hand stuffer you have, make sure you pack the meat tightly into the stuffer to avoid air spaces, as air will be pushed into the casing during the stuffing process. Turn the handle or push down the lever until the meat shows in the end of the stuffing tube. Again, make sure the casings are washed, flushed, or soaked, depending on the type used. Lubricate the "horn" or tube of the stuffer. Use water for natural casings, cooking oil for collagen casings. If making fresh sausage, brats, or other sausages using natural or collagen casings, slide the casing all the way onto the stuffing tube. Pull about two inches of casing from the tube and tie the end in a knot.

It's important to make sure the meat is well packed into the stuffer to eliminate air pockets.

Push the tied end back against the stuffing tube. This will prevent air from being trapped in the casing.

Holding the casing in place on the stuffing tube with one hand, turn the handle or push the lever with the other hand and fill the casing. Do so gently at first, until the sausage begins to fill the casing. As the sausage comes out of the tube, it will pull the casing off the tube. Allow the stuffed casings to feed out onto a large surface in front of the stuffer.

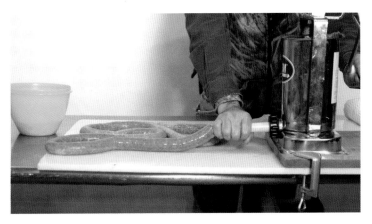

Lubricate the horn of the stuffer with water for natural casings, or use spray cooking oil for collagen casings. Slide the casing all the way onto the horn of the stuffer and tie a knot in the end.

Holding the casing in place, turn the handle or push the plunger to fill the casings. As the sausage feeds into the casing, it will force the casing off the horn. If air pockets form, prick with a pin or toothpick.

Although stuffing sausage isn't particularly hard to do, it takes a bit of practice to learn how full to stuff the casings. Two people definitely make the chore easier. If making sausages into links, it's important to not overstuff the casing. Casings will usually have some air pockets during filling. Using a sharp item such as a pin or toothpick, prick the casing to allow the air to escape. If you don't do this, the casing will burst at these locations during cooking or leave air pockets to collect mold in sausages such as summer sausage.

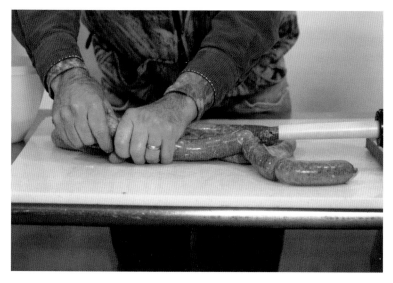

When using natural casings, twist the casings into links.

Collagen and plastic casings require you to tie off the links with butcher's string.

Fibrous casing ends are often closed with hog rings and a hog ringer.

If the sausage is to be made into links and you are using natural casings, twist the stuffed casing into 4- to 6-inch links. Simply twist four to six times to create the links. Some makers prefer to twist all in the same direction and then, after drying or smoking, they will use a sharp knife to cut through the twisted areas to separate into individual links. If using collagen casings, you will need to tie off the links, as

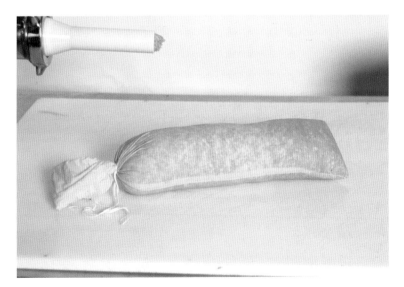

Muslin and other large casings are usually tied at the end with butcher's string.

collagen will unwind. Use cotton butcher's string tied tightly in place. Make two ties at each link space, and then cut between the ties with a sharp knife. If making a sausage such as bologna "rounds," simply tie off at the end of the casing.

If making summer sausage or others from fibrous casings, again, soak the casings in warm water for twenty to thirty minutes before using. Follow the recipe directions for mixing the ingredients, but make sure you do not overmix. Mix only until the meat feels tacky and then stuff immediately. Again, fill the stuffer with meat and pack it thoroughly in place. Turn the handle or push down the plunger until the meat begins to show at the end of the stuffing tube.

Fibrous casings usually have one end tied, or clipped, shut. Slide the fibrous casing all the way on the stuffing tube. Then hold the fibrous casing securely back in place on the stuffing tube and use the plunger or handle to force the sausage into the casing. Go gently, and make sure the tied end is tight against the stuffing tube when you begin to prevent adding air into the end of the casing. If air pockets develop, prick with a sharp object. Fill the casing to approximately one and a half inches from the end. Twist the end of the casing tightly and tie with cotton butcher's string or fasten with "hog rings" and a hog-ringer tool.

In either case, if you run out of sausage before completely filling the casing, simply leave casing in place and refill the stuffer. There will always be a remainder of sausage in the tube. If you run out of casings before sausage, simply add another casing.

④
Dry, Semidry, and Hard Sausages

HARD SAUSAGES HAVE been a tradition in many countries for centuries, primarily because some of them can be stored without refrigeration. Once cut into, however, these sausages will keep longer if refrigerated. A wide variety of hard sausages can be found. Most hard sausages today are not consumed at mealtime but used as snack or party foods. Some hard sausages may also be used in food preparation. These sausages were traditionally dried, not cooked, although cold-smoking was occasionally used to create flavor.

Because of the possibility of contracting trichinosis from the raw meat used for these uncooked sausages, freezing the meat before making the sausage is one safety precaution used by commercial producers. Freezing does not kill trichinosis in some animal meat, however, including bear, and it is also hard to do correctly at home.

These days, many of the traditionally "uncooked" sausages are thermal-heated, or cooked to an internal temperature of 160°F. One common method is smoke-cooking and then air-drying. All of these dry sausages also require the use of a curing agent, and some are also fermented.

Homemade traditional hard or dry sausages, such as summer sausage, make a great snack food to share with family and friends.

Prepared mixes make it easy to create great-tasting summer sausages.

Proper drying of these hard sausages also requires the right temperature and humidity. The ideal temperature is 45° to 50°F at about 70 to 80 percent humidity. At this temperature and humidity, it will take about sixty days to dry the sausages. This was traditionally done in the cold part of the year and, when done correctly, will remove about one-third of the weight of the sausage in moisture. The sausage will acquire some mold, but that can be washed off with a mild vinegar and water solution, or you can simply peel off the casing and mold before eating.

All of the semidry and hard sausages that are not dried must be refrigerated or frozen for future use. Dried sausages can be stored without refrigeration.

Venison summer sausage has become a favorite of ours, and we often make enough to give some away to friends and family. Summer sausage can also be made of other wild game meats, including waterfowl, as well as beef and mutton. Because of the popularity of this sausage, the nearly limitless number of ways it can be prepared, and the many recipes available, we've included recipes for several varieties.

Morton Salt Savory Summer Sausage

6 lbs. boneless pork trimmings	4 tbsps. liquid smoke
4 lbs. boneless beef trimmings	3 tbsps. sugar
1/2 cup Morton Tender Quick Mix	1 tsp. ground black pepper
or Morton Sugar Cure	1 tsp. ground ginger
(Plain) mix	1 tsp. garlic powder

This Morton Salt version is an extremely easy recipe, especially if you don't have a sausage stuffer or smoker, and it's also a great "starter" sausage project. Cut meat into 1-inch cubes and mix with remaining ingredients. Grind through a ¼-in. plate. Refrigerate overnight.

Re-grind sausage through a 1/8-inch plate. Shape sausage into slender rolls, 8 to 10 inches long and 1 ½ inches in diameter. Wrap in plastic or foil. Refrigerate overnight.

Unwrap rolls and place on broiler pan. Bake at 325°F until a meat thermometer inserted in the center of the roll reads 160°F, about fifty to sixty minutes. Store wrapped in refrigerator. Use within three to five days, or freeze for later use.

Bradley Smoker Savory Summer Sausage

5 lbs. regular ground beef (not extra lean)	1/2 tsp. garlic salt
5 tsps. curing salt	1 cup water
3 1/2 tsps. mustard seed	Cooking oil spray
2 1/2 tsps. coarse black pepper	

The Bradley Smoker folks offer this great recipe for use with their electric smokers. For simplicity, it uses purchased ground hamburger meat; however, venison or part venison could be used. In a large glass bowl, mix together all the ingredients by hand. Cover with plastic wrap and refrigerate for twenty-four hours, mixing twice during this time. Form the meat into five rolls, approximately 3 inches in diameter. Place the rolls on oiled smoker racks.

Most sausages use a combination of fat and lean meat. Grind fat separately.

Shown are bowls of ground pork fat and venison, ready to be made into sausage.

Preheat the Bradley Smoker to between 200° and 220°F. Place rolls on racks in the Bradley Smoker, and, using Bradley hickory-flavor bisquettes, smoke/cook for approximately four to five hours until a meat thermometer reads 160° to 170°F.

To store, wrap in foil, put in plastic bags, and freeze. Take out an hour before serving, slice, and serve with cheese and crackers as an appetizer.

Venison Summer Sausage

3 lbs. venison, elk, or buffalo	½ tsp. ground ginger
2 lbs. pork trimmings	½ tsp. ground mustard
1 tbsp. black pepper	1 tsp. garlic powder
5 tbsps. Morton Tender Quick Mix or	½ tsp. ground red pepper
Morton Sugar Cure (Plain) Mix	4 tbsps. corn syrup
1 tsp. ground coriander	½ tsp. liquid smoke

Weigh meats separately. Cut chilled meat into 1-inch cubes. Grind through a 3/16-inch grinder plate. Thoroughly mix ingredients in a glass bowl and pour over ground meat. Mix meat and ingredients. Place in plastic or glass bowl and refrigerate overnight.

Spread meat out to about 1-inch depth in a shallow, flat pan and place in a freezer for an hour or so or until the meat is partially frozen.

Weigh the meat and fat separately to achieve the correct proportions.

Remove the partially frozen meat and re-grind through a 1/8-inch plate. Stuff the ground meat into synthetic casings. (If you have a meat grinder with stuffing attachment, you can do this in one step.)

Hang the stuffed meat on drying racks and dry at room temperature for four to five hours or place in a smoker on sticks, with the damper open and no fire or heat. Allow to dry until casings are dry to the touch. Raise the temperature of the smoker to 120° or 130°F, add smoke chips, and smoke for three to four hours. Raise the temperature to 170°F and cook until the internal temperature reaches 165°F. Shower with cold water. Place sausages back in the cooled-down smoker and hang at room temperature for one to three hours to dry and bloom the sausage.

Peppered Beef Summer Sausage

3 lbs. regular beef (not lean)	½ tsp. ground red pepper
2 lbs. pork trimmings	1 tbsp. garlic powder or 5 minced garlic
5 tbsps. curing salt	cloves
1 ½ tsps. fresh ground black peppercorns	1 cup cold water
½ tsp. uncrushed peppercorns	

If you like the taste of black pepper, this version will be a favorite. Weigh meats and grind using a 3/16-inch plate. Mix spices and pour over ground meat. Blend thoroughly. Place in a covered plastic or glass bowl and refrigerate overnight.

Mix well, roll into logs logs 1 ½ x 12 inches long, and wrap in plastic or foil or stuff in 2 ½-inch fibrous casings and refrigerate overnight again. Cook on oven racks at 325°F or smoke at 130°F for two hours. Raise temperature to 160°F and smoke for two more hours. Raise the temperature to 180°F and cook until internal temperature reaches 160°F.

Lamb Summer Sausage

3 lbs. lamb	1 tsp. marjoram
2 lbs. pork trimmings	1 tbsp. sugar
5 tbsps. Morton Tender Quick Mix or other cure	1 tbsp. garlic powder or 5 minced garlic cloves
1 tbsp. black pepper	1 cup cold water
1 tsp. mustard seed	

A mild-flavored summer sausage with the distinct flavor and texture of lamb, this sausage will go well with stronger-flavored cheeses. Weigh and coarse-grind the meats. Mix in the cure. Place in a shallow pan and place in a cooler or refrigerator for three to five days. Remove, add the spices, and re-grind using a 1/8-inch plate. Partial freezing will help with this task. Stuff the sausages into 2 ½-inch fibrous casings.

Place in smoker and smoke at 130°F for two hours. Raise temperature to 160°F and smoke for two to three hours. Raise temperature to 170°F and cook until internal temperature reaches 160° or 170°F.

Easy-Does-It Deer-Season Venison Sausage

3 lbs. venison	5 tbsps. curing salt
2 lbs. pork trimmings	1 tsp. black pepper
1 tbsp. mustard seed	1 cup cold water
1 tsp. liquid smoke	

This is a quick and easy sausage to make when you first bring home the venison. Bone off the venison and cut the meat into 1-inch cubes. Refrigerate overnight. Cut the pork into cubes and weigh both meats. Grind both through a 3/16-inch plate. Mix the cure and spices in a glass bowl and pour over the meat. Mix the meat and spices well, adding the water slowly as you mix. Shape the sausage into rolls and wrap each roll in plastic wrap. Refrigerate overnight. Unwrap the rolls and place on broiler pan. Bake at 325°F until internal temperature reaches 160°F. Store wrapped in the refrigerator for a couple of days, and freeze for longer storage.

Cervelat Summer Sausage

2 lbs. beef chuck	5 tbsps. curing salt
2 lbs. fatty pork butt or pork trimmings	1 ½ tsps. ground coriander
1 lb. beef or pork hearts	1 tbsp. garlic powder
1 tsp. ground black pepper	1 tsp. ground mustard

Cervelat sausage is an extremely popular German sausage that is also popular in Sweden and normally consists of more pork than beef. Larger pieces of pork fat in this recipe make the appearance of this sausage distinctly different from that of other varieties.

Cut the beef chuck and hearts into 1-inch cubes and grind through a 1/8-inch grinder plate. Cut the pork pieces into 1-inch cubes and grind through a 3/16- or 1/4-inch plate. Place the meats and seasonings in a tub and mix thoroughly. Stuff the meat into 2 ½-inch synthetic or fibrous casings and place in the refrigerator for forty-eight hours.

Hang the meat on drying racks and dry at room temperature for four to five hours, or place in a smoker on sticks with the damper open until the casings are dry. Raise the temperature of the smoker to 120° to 130°F, add smoke chips, and smoke for three to four hours. Raise the temperature to 170°F and cook until the internal temperature reaches 165°F. Shower with cold water. Place the sausages back in the cooled-down smoker and hang at room temperature for one to three hours to dry and bloom the sausage. This sausage is a good choice for drying after cooking.

Jalapeño-Cheese Venison Summer

3 lbs. venison	1 tsp. marjoram
2 lbs. pork trimmings	1 tbsp. sugar
5 tbsps. cure	1 tbsp. garlic powder or 5 minced garlic
1 tbsp. black pepper	cloves
1 tsp. mustard seed	½ lb. high-temperature jalapeño cheese

If you like a bit of heat with your summer sausage, you'll enjoy both the heat and the tasty addition of cheese. This is a very popular sausage at our deer camps. The cheese must be a high-temperature cheese that will not melt at up to 400°F, so you can smoke and cook the sausage as you would normally. Weigh the meat and cut it into one-inch chunks. Mix the cure and seasonings in a glass or plastic bowl

Regardless of whether dried, semidried, or hard sausages, they should all be cooked to an internal temperature of at least 160°F.

Want a little variety? Try jalapeño-cheese venison summer sausage.

and sprinkle over the meat chunks. Mix the meat and seasonings well. Grind the seasoned meat through a 3/16-inch plate. Place the ground meat in a large bowl or tub and add the cheese, making sure it is well distributed throughout the ground meat. Stuff the meat and cheese into 2 ½-inch fibrous casings.

Special high-temperature cheese is added to the ground meat.

After cooking, sausages should be immediately immersed in ice water to cool them down quickly.

Hang the sausages on drying racks and dry at room temperature for four to five hours, or place in a smoker on sticks, with the damper open, until the casings are dry. Raise the temperature of the smoker to between 120° and 130°F, add smoke chips, and smoke for three to four hours. Raise the temperature to 170°F and cook until the internal temperature reaches 165°F. Shower with cold water. Place back in the cooled-down smoker and hang at room temperature for one to three hours to dry and bloom the sausage.

Beef Salami

1 lb. beef 1
lb. beef chuck
1 ½ level teaspoons Morton Tender Quick Mix or Morton Sugar Cure (Plain) Mix
1 tsp. Morton table salt
(For a spicier version, substitute 1 ½ tsps. Morton Sausage and Meat Loaf Seasoning Mix.)

½ tsp. mustard seeds
½ tsp. freshly ground black pepper
½ tsp. garlic powder
1/8 tsp. nutmeg
Few drops of liquid smoke, if desired

An easy way to make summer sausage is to simply roll the ground and seasoned meat into logs.

From the Morton Salt folks, this cooked salami is extremely easy to prepare, and the small amount provides a chance to try your hand at this type of sausage making.

Combine all ingredients, mixing until thoroughly blended. Divide in half. Shape each half into slender rolls about 1 ½ inches in diameter. Wrap in plastic or foil. Refrigerate overnight.

Unwrap. Bake on a broiler pan at 325°F until a meat thermometer inserted in the center of a roll reads 160°F, about fifty to sixty minutes. Store wrapped in refrigerator. Use within three to five days, or freeze for later use.

Wrap the rolls in plastic wrap and refrigerate overnight to meld flavors and set the sausages.

Smoked Salami

3 lbs. lean chuck beef	1 tbsp. garlic powder or 5 minced garlic
2 lbs. fatty pork	cloves
3 tbsps. soy protein concentrate	5 tbsps. curing salt
3 tbsps. corn syrup	1 tbsp. cardamom
1 tbsp. ground black pepper	1 cup ice water
1 ½ tsps. whole black pepper	

This is excellent smoked salami that also is cooked in the smoker. The larger pork pieces add to the appearance and flavor. Weigh the meats and keep separate. Grind the pork through a ¼-inch or coarse plate and the beef through a 1/8-inch or fine plate. Mix the meats together and add the other ingredients. Mix well and stuff into 3 ½-inch fibrous or cellulose casings. Refrigerate the sausages overnight.

Hang the meat on drying racks and dry at room temperature for four to five hours, or place in smoker on sticks with damper open until the casings are dry. Raise the temperature of the smoker to between 120° and 130°F, add smoke chips, and smoke for three to four hours.

Unwrap and bake in a 325° oven until the internal temperature of the sausages reaches 160°F.

Raise the temperature to 170°F and cook until the internal temperature reaches 165°F. Shower with cold water. Place back in the cooled-down smoker and hang at room temperature for one to three hours to dry and bloom the sausage. This sausage can also be dried after cooking.

Cooked or Smoked Salami

3 lbs. lean beef chuck	½ tsp. nutmeg
2 lbs. pork shoulder butt	½ tsp. allspice
½ tsp. Bradley Sugar Cure	¼ tsp. ground red pepper
1 tsp. black peppercorns, cracked	¼ cup sherry (optional but recommended)
2 tsps. paprika	2 tbsps. light corn syrup
1 tsp. ground black pepper	½ cup water
1 tsp. onion powder	1 cup finely powdered dry skim milk
1 tsp. garlic powder	

From the Bradley Smoker folks, this is great smoked salami with lots of flavor. Soak fibrous casings in water for thirty minutes prior to using. Four casings will be required if they are 2 ½ inches in diameter and about 12 inches long. Grind the beef and pork through a 3/16-inch plate.

Mix the seasonings, cure, water, and powdered milk in a large bowl until the ingredients are perfectly blended. (For a normal salt taste, add the optional 1 teaspoon of salt; for a mild taste, omit the salt.) Add the meat to the mixture and mix thoroughly. Knead for about three minutes.

Stuff the sausage mixture into the fibrous casings. Insert the cable probe of an electronic thermometer in the open end of one of the sausages. Close the casing around the probe with butcher's string. Refrigerate the salami overnight.

Remove the sausage from the refrigerator, and place it in a smoker that has been heated to 150°F. Make sure the damper is fully open while drying the surface of the casings. Maintain this temperature with no smoke until the casing is dry to the touch. (Alternatively, dry the casings in front of an electric fan.) Raise the smoker temperature to 160°F, add the Bradley wood bisquettes, and smoke for three to six

Some "hard" sausages, such as salami and pepperoni, are traditionally dried under the right conditions. The Sausage Maker, Inc., has a complete kit for creating these traditional sausages.

Although many hard sausages are stuffed into casings and smoked or dried, there are also easier ways to make these kinds of sausages (including pepperoni).

hours. If you wish to cook the sausage in the smoker, raise the temperature to 180°F and cook until the internal temperature is 160°F.

Instead of final cooking in the smoker, the sausages may be cooked by steaming. After smoking the sausages for three to six hours, wrap each sausage in plastic food wrap (optional), and steam until the internal temperature is 160°F. A steamer may be improvised by using a large pan with an elevated rack inside, covered with a lid.

As soon as the cooking is finished, chill the sausage in cold water until the internal temperature drops below 100°F. Refrigerate overnight before using.

Morton Salt Pepperoni

1 lb. lean ground beef (chuck)	½ tsp. mustard seed
1 ½ level teaspoons Morton Tender Quick Mix or Morton Sugar Cure (Plain) Mix	½ tsp. fennel seed, slightly crushed
	¼ tsp. crushed red pepper
1 tsp. liquid smoke	¼ tsp. anise seed
¾ tsp. freshly ground black pepper	¼ tsp. garlic powder

Homemade pizzas taste even better with pepperoni sausage you've made yourself. The small batch in this recipe, from Morton Salt, also provides a quick and easy way of creating a semidry "hard sausage."

Grind beef through a 3/16-inch plate. Combine all ingredients, mixing until thoroughly blended. Divide mixture in half. Shape each half into a slender roll about 1 ½ inches in diameter. Wrap in plastic or foil. Refrigerate overnight. Unwrap rolls and place on a broiler pan. Bake at 325°F until a meat thermometer inserted in the center of the roll reads 160°F, about fifty to sixty minutes. Store wrapped in the refrigerator. Use within three to five days, or freeze for later use.

Bradley Smoker Wild Game Pastrami Sausage

4 lbs. wild game meat (venison, elk, moose)	½ tsp. Bradley Sugar Cure (do not use more than this amount)
1 lb. pork fat	1 tsp. salt (optional)

4 tsps. light corn syrup	½ tsp. oregano
4 tsps. black peppercorns, cracked	¼ tsp. allspice
1 tsp. onion powder	¼ tsp. ginger powder
1 tsp. garlic powder	½ cup water
½ tsp. cayenne	1 cup fine powdered skim milk
½ tsp. paprika	

Using a combination of smoking with the Bradley Smoker and cooking in an oven, this is an easy and spicy pastrami recipe. Soak 2 ½-inch fibrous casings for thirty minutes before stuffing. Grind the meats through a 3/16-inch plate.

Mix the seasonings, cure, water, and powdered milk in a large bowl until the ingredients are perfectly blended. (For a normal salt taste, add the optional 1 teaspoon of salt; for a mild salt taste, omit the salt.) Add the meat to the mixture and mix thoroughly. Knead for about three minutes.

Stuff the sausage mixture into the fibrous casings. Insert the cable probe of an electronic thermometer in the open end of one of the sausages. Close the casing around the probe with butcher's string. Refrigerate the salami overnight.

Remove the sausage from the refrigerator and place it in a smoker heated to 150°F. Make sure the damper is fully open while drying the surface of the casings. Maintain this temperature with no smoke until the casing is dry to the touch. (Alternatively, dry the casings in front of an electric fan.)

Raise the smoker temperature to 160°F, add the Bradley wood bisquettes, and smoke for three to six hours. If you wish to cook the sausage in the smoker, raise the temperature to 180°F and cook until the internal temperature is 160°F.

Instead of final cooking in the smoker, the sausages may be cooked by steaming. After smoking for three to six hours, wrap each sausage in plastic food wrap (optional), and secure the ends with a wire bread-bag tie. Steam the sausages until the internal temperature is 160°F. A steamer may be improvised by using a large pan with an elevated rack inside, covered with a lid.

Many hard sausages are also smoked for flavor or hot-smoked for cooking.

As soon as the cooking is finished, chill the sausage in cold water until the internal temperature drops below 100°F. Refrigerate overnight before using.

Dried Sausage Sticks (Slim Jims)

1 lb. venison	¼ tsp. garlic powder
1 tbsp. Morton Tender Quick Mix	¼ tsp. Worcestershire sauce
¼ tsp. black pepper	1 tsp. crushed red pepper
½ tsp. onion powder	1 tsp. liquid smoke

My nephew Morgan created this great jerky-style sausage recipe that can be made in small batches. Simply take a pound of previously ground venison or other red meat (burger) out of the freezer, thaw, and mix in the ingredients. Cover with a plastic wrap in a plastic or glass bowl and refrigerate overnight. The liquid smoke isn't added to the ground meat; instead, it's sprayed on during the drying process.

To make the Slim Jims, extrude the ground and seasoned meat through a jerky gun (using the small round nozzle) onto oiled cookie

Premixed seasonings, such as the pepperoni mix from Bass Pro Shops, make it easy to make your own Slim Jims.

Use a jerky gun or, in this case, an extruder on the LEM/Bass Pro grinder to easily extrude the sausage sticks.

sheets, or a jerky screen for oven drying, or dehydrator trays for drying. Morgan dries the batch long enough to set the top side, then sprays that side with liquid smoke.

Halfway through the drying process, he turns the strips over on the dehydrator trays and sprays the other side. Morgan says the liquid smoke isn't lost in the ground-meat mixture, and the sausage has a fresher, smoked flavor. You can also brush on liquid smoke with a pastry brush. The meat must be dried/cooked to an internal temperature of 160°F.